A Shadow in the City

CHARLES BOWDEN

A Shadow in the City

Confessions of an Undercover Drug Warrior

A Harvest Book

HARCOURT, INC.

Orlando Austin New York San Diego Toronto London

Requests for permission to make copies of any part of the work should be submitted
online at www.harcourt.com/contact or mailed to the following address: Permissions
Department, Harcourt, Inc., 6277 Sea Harbor Drive, Orlando, Florida 32887-6777.

www.HarcourtBooks.com

The Library of Congress has cataloged the hardcover edition as follows:
Bowden, Charles, 1945–
A shadow in the city: confessions of an undercover drug warrior/Charles Bowden.—
1st ed.
p.   cm.
1. Narcotic enforcement agents—United States.   2. Narcotics, control of—United
States.   3. Undercover operations—United States.   I. Title.
HV5825.S695 2005
363.45'0973—dc22   2004026622
ISBN-13: 978-0-15-101183-4   ISBN-10: 0-15-101183-4
ISBN-13: 978-0-15-603253-7 (pbk.)   ISBN-10: 0-15-603253-8 (pbk.)

Text set in Minion
Designed by Linda Lockowitz

Printed in the United States of America

First Harvest edition 2006
K J I H G F E D C B A

an old preacher holds on to a bible
a small child sings the water is wide
I stand like a stone all alone outside their circle
a faithless man, a fast left hand and a hole inside
—Ray Wylie Hubbard, "the sun also rises"

I T ALL HAPPENED. Everything takes place in an unnamed city that exists. The man who calls himself Joey O'Shay continues his life. But that is not his real name. Nor are the other names real. Nothing has been fabricated, no person is a composite. The deals occurred. As did the killings, beatings, shootings, tortures, betrayals, suicide, and love.

Joey O'Shay says, "She gripes at me. I ponder the situation having had a few. I tell her undercover work...working heroin with old Latinos...is like playing poker in a dingy old yellowish lighted 1920s hotel room...with a oscillating antique fan...with ghosts. She starts laughing. I tell her we all have ghosts...but they bring them hard to the table. No wild cards...straight poker. Old men including me...dragging our past to play. No one pissed off...just playing what we're dealt. Nobody goes home... both me and the ghosts...we know it."

For my sons,
Joey, Rocky, Skipper
The honor of being your Father,
K.S. (Joey O'Shay)

Deep Night

hadow n the ity

1

HE PUTS THE GUN DOWN. It is close, his finger beside the trigger, the round chambered and ready. Then the man moving toward him hesitates, and the moment passes. Joey O'Shay's face remains passive, a blank as he reaches for the gun and sweeps it up, a blank as he lowers it and puts it down.

He says, "Never hesitate, never say a fucking thing, none of this hands-up shit. Just keep pumping them into him. If you want to live."

O'Shay looks at the departing man without expression.

The bluing has worn off the weapon.

The city does not sleep at this hour but becomes a zone of zombies, that time when the drunks have made it home, the sober have not yet risen, and the streets belong to the feral, to predators coursing its arteries for prey.

Deep night.

He cannot rest. The nights have always been hard but now with this major heroin deal bubbling along, he cannot sleep at all. When he was younger, he prided himself on not needing sleep. Now he is resigned to not getting sleep. He thinks of $5 million a week street value and then he does not sleep at all. Not out of greed but because of other hungers.

"One of my sons wrote this about me," he says. Then he slides the music into the dash of his truck.

A voice knifes through the stale air:

> Come take a walk inside my mind,
> Meet the ghost that lives inside
> Fallen friends and broken dreams
> That haunt me in my sleep.

COME, SMELL, LOOK, LISTEN. There, the city spreads like an oil slick across the flat plain at the point where the rain begins to die. The past falls dead here before it is born—bladed, buried, and unmissed. Yesterday has no more meaning than last night's bar bill.

Everyone comes here for the money and no one knows why the city itself exists. The sky refuses to forgive, the sun seldom smiles. The air sags with fumes and towers rise and try to center the glistening slick with no tool but money. The talk is boasts, the lips thin, the streets a refutation of the talk. As in all such places, most people believe in the promise and most lose. In the dirt and grass and wind and tired rivers that lie beyond the city, the eyes tighten at its name and one and all say they hate the city. And yet they come regardless of their hatred. And so the city thrives and devours despite the words of anyone. Streams barely move across the flat earth and the waters laze and eddy with trash and time.

Smell the exhaust of millions laid out like a pure line for all to suck up through rolled hundred-dollar bills. The grass rank by the ditch, the green water licking the air, diesel fumes foaming out of the trucks, a woman walking briskly in heels and trailing musk, piss in the alleys, raw onion chopped and biting from the small food stalls, pecan smoke reaching out from under the brisket, ribs, and sausage, the fresh tang magically lifting from the beaten streets after a sudden spring shower, the smell of a child's hair, that scentless scent shared with fawns hiding in the tall grass, the faint scent of a child's hair slamming the face in the dark of the night. The gunpowder shredding the stale air and hanging there—a noose and a gallows after the explosion has passed.

Look at the spires scratching the sky, squares, rectangles, spikes, all insulting the sun and the moon and the stars, houses hiding on the prairie, faces blank and safe, white, black, brown, vacant, nothing in the eyes but the city staring back, eyes careful, eyes eager, eyes always alert, eyes never trusting, the towers a gleam on the corneas, towers beyond reach, towers saying into the day and the night a yahoo and yodel to the prairie that fails around them. A woman's body rigid at a table as the waiter serves, his jowls sagging as he takes her in, a man leaning into the window and whispering fast words in code and his hand reaching for the hundred-dollar bill for the message, the idle of engines always at red lights never at green, the hope of Saturday night melding into the pollution of Sunday morning, always, always coming down, streets not mean but cold like the mortician who screws the coffin lid shut. Screams, laughs, alarms, wails, bottleneck strumming, quick picking, blue haze of a bar when everything briefly feels right and beckoning, soft music at two a.m. in the dark with the drink warming in the hand, whisk of tires down midnight streets, light creak of black leather as she walks her tight pants past, the thrumming of the fingers on the cleared desktop, the maps and plots where the money lies hidden in the ground, the lockstep as people move from their cages to their prisons, the towers rising and rising and saying join us or die.

Listen as the air brakes jack the ear on the big roads lacing the city and moving its blood like sludge, hear the horns, the choppers—whomp whomp whomp—overhead on their secret errands, the shout of children racing through the back lanes, the chirp and crackle of birds stalking the crumbs and garbage, the click of keys in the towers, the hum of overhead lights in the caves of work, the soft rich vowels of Spanish in the back rooms of businesses, the chords of a blues guitar asking for someone to consider the question, long sigh of a zipper down the back before the dress melts to the floor, the bark of angry dogs, the slippery song of knife sliding into flesh, blade warming itself with blood, the lights at night fighting the prairie, beams, shreds, slabs, towers, beacons of light that only seem to underscore the loneliness as people pull the shades in their houses and lock doors and scurry down darkened walkways and pray for dawn. The light golden in the fat wet air or glaring through the breath off the parched plains to the west, the choirs faltering toward heaven from the temples of Sunday, thunk of a shot glass after that necessary swallow, voices loud, braying, vowels licked and slurred, consonants like ice picks, voices clamoring for attention as the machines smother them with decibels, thrash of the tree limbs the night of the big wind, sirens, chimes, radios, televisions, bar bands, lap-dancing palaces, a singer saying the city "is a rich man with a death wish in his eyes," a preacher saying the city "is lost but must be saved," slap of shoes on dark streets, the audible click of eyes as the young men with guns lock onto a target, the silent prayer of her breasts falling from her bra, the faint promise, barely a whisper, of lipstick spreading on the lips, the shout of a hammer locking on a gun.

Stories begging and getting not a dime, stories never written, never sung, stories from the place where the light does not go and where the pages refuse to turn, stories without endings or beginnings, simple stories like the city itself.

A small creek laces through the shotgun houses, the banks a maze of trees and canebrakes. Long ago, a humpbacked form slides silently down the creek, a huge beast with a hard back moving and hunting and yet the city knows nothing of this beast and of the beast's habits and ways. It is as if the violent appetite slipping down the calm waters did not exist and had no past or present or future. A boy watches at dusk from the bank and never forgets that moment or the feel of that moment.

He becomes a shadow in the city.

He remains unknown to the life moving around him.

He answers no questions.

He grows, thrives, slides silently down the streets.

He acts.

He loves.

He loses.

He kills.

He is the law but few remember this fact.

Sometimes, he forgets himself.

BOBBIE DOES NOT LIKE THIS DREAM. She's been in his deals for twenty years, but she's never let Joey O'Shay into a dream before.

In the dream, the phone rings and it is Joey. He tells her to go on an errand and so she does, she always does what he tells her to do. Bobbie obeys no one, she is proud of this fact. But she always obeys Joey. Even in her dreams.

When she gets to the place, she discovers it is a trap and she cannot escape. So she calls Joey and he says, "I'll send a car to get you."

And he does. He always does what he says. He always is in command.

As Bobbie steps into the car, two things happen. First she feels a stab of pain in her back and she thinks, My God, someone has shot me. And then she looks out the car window and sees Joey O'Shay standing across the street, and he is very safe and calm in the shadows of the city at night.

None of this, of course, matters to the actual deal. The plane is coming in, the money ready. The heroin is warehoused and waiting to be delivered.

It is 94 percent pure, the best Colombia has to offer. And there is no limit on the amount.

Heroin is the future. No one walks away from heroin. It is the love of life itself. And all the true lovers gather around its soft and clean whiteness. Heroin does not lie, does not leave you alone for a single instant. Even when heroin walks out the door, soon the slap of shoes hits the pavement, chasing heroin down the street. Such is love when it can be found.

JOEY O'SHAY'S SHOULDERS are powerful from lifting weights. His hands are strong, his feel for flesh instinctive. The trick is never to hesitate, not for a fraction of a second. Act and let God sort it out.

He and one of his gunmen are in the dark when the other man flinches. Something is back there in the dark, a hulking thing.

O'Shay says softly, "You saw it, right?"

The man with the gun says, "Yes."

A calm settles over O'Shay, his solid body relaxes briefly. He feels an odd sensation, he feels for a flickering moment that he is not absolutely alone because someone else has seen the blackness on his trail.

The thing that follows him is not listed in any field studies. No tracks have been discovered, no spoor has been found. But O'Shay can describe it exactly.

...the thing is big and dark, a hulking form that somehow moves on cat's paws and casts no shadow, makes no sound, not a single fucking word, but never stops coming, follows me down all the streets and alleys. A filthy fucking thing. The Evil Creature...

Whenever O'Shay feels it near, he is less alone in this world of words and lies. He is close to something pure. Like heroin, a clean and absolute thing. Then he truly relaxes.

FOR MORE THAN TWENTY YEARS he has lived underground, a predator feeding off the flesh of the city. He began working a neighborhood, then he entered the commerce of flesh, and finally found his place in the drug world. He discovered he was better than good at the work. Soon he was pulled into a group that melded local people with national talent. He is a person of no particular standing. He wears street clothes. In a hive of agents and bureaucrats, he operates independently. This is permitted because he makes deals and others can ride on the success of these deals. This suits him because he does not like being told what to do. He's killed, been in shootings, gone into rooms too many times where guns began firing, but he has never been shot. But he cannot speak of these matters. There is an etiquette in his world that holds such talk obscene, a kind of bragging left to liars and to those with no morals. He can barely use the word "kill." Not because of a need for euphemisms but because of the demands of silence. To take a life is bad, to speak of it unforgivable.

For years, O'Shay has had a goal, one he keeps to himself, one he resists and yet cannot really put aside. This goal can be felt but not said. This goal is almost a whisper in his head: he wants someone to end it for him. He is very intelligent and he has assessed his situation carefully. He lacks the appetite for suicide. Nor can he fail to give his best in any situation. Try as he may, he cannot lose that half step, he cannot dim his alertness, he has repeatedly failed to silence his instincts. He cannot stop himself. He has been in all those firefights, but he has never been wounded. He feels he is cursed.

He has done major marijuana operations. He has done major cocaine deals. He has gone into rooms repeatedly with nothing but his wits and his gun, with backup ready but useless once the violence begins because the violence comes without warning and when it sweeps through the room, he is on his own for those twenty seconds, those thirty seconds, sometimes that minute, alone with his gun and his fate. He has stood in a room where a barrage filled the air and still he has not been shot, much less killed.

There are conditions that must be met if he is to die and end this thing.

He must be beaten by someone intelligent. He must be finished by someone cold, a fellow predator. He must meet someone who is better at killing than Joey O'Shay.

And there is one final condition: he does not want to die.

That is how the deal begins.

Heroin attracts the best people, the smartest, the coldest, the deadliest.

In this world of heroin, the Colombians are reputed to be the best. They are honest, efficient, and savage. Failure means death not simply for the individual but for his women and children, for his mother and father. And yet the Colombians have an integrity that O'Shay admires. They work hard, they keep their word, they are punctilious about business obligations.

O'Shay hates anyone without a sound work ethic. He cannot abide many cops for this very reason. He cannot accept modern DEA agents and the other nine-to-five cops. There have been moments in the office when he has cussed and threatened such people.

That is how the deal begins one fine April day as Joey O'Shay sips iced tea in his city, a city that is unaware of his existence. In more than twenty years of violence, his name remains unknown. O'Shay hates publicity, not because it would threaten his work, though it surely would, but because it violates that work, it makes the work dirty and self-indulgent. And for O'Shay the work is sacred and must be kept clean and pure. That is why he does not care who gets credit for his deals.

He refuses business with small operators. He will not stir for small loads. He has contempt for work built on loose talk, allegations, indiscretions. For him, there must be drugs and money on the table and the deal must mean serious weight. Otherwise he is corrupting the work. He will not foul his destiny with shortcuts and pettiness.

He will not speak of the dead. That would be disrespectful. He shares a bond with his dead, a bond that takes shape in the air and the scent and the intimacy that envelops two human beings so close and so focused on each other that they fuse into one identity, in the brief interlude before they become the slayer and the slain.

He is almost fifty years old. He is tired, bone tired, and not from the nights or the work or the endless hours.

So he will do a major heroin deal, he will reach out to Colombians. He will test himself against the very best. He will see if they will stop him. Before he is done with it, he will have reached into cities on three continents, he will have penetrated realms unknown to the authorities who sit at their desks fiddling with budgets and cases. He will have created an economic hole that swallows tens of millions of dollars. He will have ruined hundreds of human beings, and brought unknown others under torture. And he will have left his underground world, either as a corpse or as a living and breathing man.

All this he decides over iced tea in a sleepy midday café.

There is a moment when one can kill without even spilling the tumbler in one's hand. A cold moment.

There is a moment when one can weep without warning or apparent reason. A warm moment.

Joey O'Shay is in both moments and they now spread out until they are time itself.

———

BOBBIE IS WORRIED. Not simply by the dream. She is worried that O'Shay is losing his judgment. Twenty years of deals with O'Shay, but now for the first time she is worried about him. He talks about quitting, about one last deal, about being sick of the business. She cannot tolerate this talk. It is not like him, she thinks.

With a knife and fork she delicately clips the tails off the fried shrimp before her. The light in the restaurant is clear yet soft on the hardwood table. Beads of moisture sweat on her water glass, her skin looks ghostly against all the dark wood. She is buxom, clear-eyed. She moves like a woman who has had many men want her, the slow ease of knowing they will come when beckoned. She dips the shrimp into a red sauce, her face a perfect blank as she chews the morsel. She considers O'Shay the perfect deal maker. He is obsessive about planning, he thinks of everything. She thinks that he is addicted to risk and violence, that he will never quit but must be killed.

"Look," she says, "we were doing this deal—cocaine, huge amounts of heroin—and Joey has set up the big meeting, the one where the money and dope exchange hands. The one where you know you can get killed. He's got three guys down the hall in a room with all the monitors of the video cameras and the tape machines. In a tower across the parking lot, he's got some agents with rifles who can see into the room. And God, Joey's running around like some girl getting ready for her prom date, you know, he's checking all the video cameras, making sure the sound is clear and good, briefing all the other agents on their roles. He's frantic and yet calm, that focused way he gets. He's spent months on this deal, he never hurries, you know, he knows nobody real in this business hurries, just dumb cops who show their hand by being anxious.

"Joey fusses over each detail—that camera, is it working

right? All this stuff. But what I notice is this: nobody will be able to save him if things go bad. None of his snipers or agents down the hall, none of this video stuff will matter if someone pulls a gun. Because I time it and no matter how fast anyone races down that hall in the hotel to get to the room where Joey is doing the deal, well, I don't care if they are Olympic athletes, they're going to burn twenty to thirty seconds before they can burst through the door and fire. And he's going to be dead.

"So there it is. Months of planning, endless meet ups with the local heroin guys, haggles over price, the whole act, and now this wired room and it comes down to him being alone and no one can save him. And what's he worried about? All this shit detail for evidence. Just as he's piled up endless videotapes of earlier meetings with the heroin boys. He never makes a misstep, shit, even I can't believe he's a cop. And the moment comes, he meets with the guys in the room, and the fussing stops. He's calm, he's in that place, that goddamn zone he goes to. He's ice. And he isn't acting."

Bobbie married once and had a child, and then the child died.

"And," she explains, "I never felt pain like that before, never, and I thought I would die. I got a divorce, came out of it, well, never really came out of it, but I did get better. And I decided that I would never hurt like that again, not ever, that I'd make sure, I'd never let anyone get that close to me again. That would fix it. A simple plan.

"And I never have. Problem solved."

She laughs, a whip crack of sound, and flashes a smile. She pulls out a cigarette, inhales deeply, lets the smoke slowly trail out and spill onto her dark blue suit and the bodice of her sheer blouse. The pale flesh of her breasts glows dimly through the fine fabric, around her neck is a silver choker.

When things went bad years ago with that killing of one of

his partners, she sat with O'Shay in a car for four hours as he worked things out. He swore then he would quit, that nothing in the business was worth what he was feeling at that moment. He was finished, fuck the money, he could always take care of his needs, he would be fine. Four hours sitting with him in a car while he debriefed himself and made his vow. One he soon broke, but still he sought her out at the moment and shared those hours with her alone.

She knows him and he knows her.

So why is he suddenly in her dream?

FIRST, TURN OFF THE EXPRESSWAY, then down a paved side road to a building that looks like a bunker. The building has no sign announcing its function. Cyclone fencing topped with razor wire encircles the structure. At the gate a tire-shredding device blocks entry, the guard must see credentials. At the entrance to the underground parking garage another machine demands a coded card before the door opens. The motor dies in the coolness of the garage. The stairs are lonely as feet plod up to the third floor. Through the door the gray face of federal offices comes into view, the bland carpet, the fluorescent lights, the dropped-tile ceiling. The air is stale, machine air fed by vents in a building where no windows ever open. The scent of the place is also sterile, the low reek of cleaning fluids and dust particles. Credenzas form workstations that sprawl like cells in some dull organism, the clutter of lives—family pictures, favored baseball caps, souvenirs of travels—lining their shelves. At the far end of this haven for officers, this cave where the task force toils, a giant U of workstations stands empty save two. One is the desk where Joey O'Shay works his phones and stage-manages his little theater of deals. The other desk is different.

A skull rests on a column. A cross is topped by a carved

African head. Seashells, coins from Mexico and Jamaica, and three 9-mm rounds with brilliant copper tips lie scattered. There are bones from fish and birds, figurines from Central America, cigars, more candles, a monkey carved from a coconut. Two clear bottles hold faces cut out from photographs. These are doomed.

Beneath the altar and flickering candles is a wooden box painted flat black and shaped like a coffin. A slit is cut in the top. O'Shay made the box with his own hands. He puts people there when he is finished with them.

He says at such moments, "They've been bad," and a smile barely flicks across his lips.

The blue fly in the bottle is kept under lock and key.

This is the altar maintained by Joey O'Shay. It is in plain view but no one asks him about it. Or about the changes he makes in the altar from time to time. And no one comes near it. Or him.

O'SHAY REACHES OUT through his network, the small army of criminals he has fed and nurtured for years.

He calls Cosima and explains what he needs. She has the right phone numbers. In her time, she has moved tons of cocaine, she has had men murdered. For one of those episodes, she went to prison. When she was released, she was extradited. Now she moves around the city like a cat, with contracts on her life trailing her. When he met Cosima, she had this problem: a local hit guy was robbing her. O'Shay listened, picked up the phone, and said, "Do me a favor: stop robbing this bitch." And so it ended, and his relationship with Cosima began. Officially, she is a CI, a confidential informant who has given up her grifting ways to work for the government and the betterment of the United States. This fiction is legally essential, and O'Shay does not question her about her other activities. He is the one who ironed out her return to the United States. One slipup, and he can have her deported. Or perhaps jailed. But O'Shay is only interested in the work. What Cosima does with her spare time is not worth his bother.

When she was younger she used to fuck Carlos Lehder when he was operating in the Caribbean. She has Italian mafia connections, links to South Africa, to the Russians, to the Arab world. Soon after prison and her return from Mexico, Cosima was making deals with people in prison, both in the United States and Mexico, she was cutting deals with the Mexican cartel, was getting shipments out of Los Angeles. Cosima became the person who could do anything for O'Shay. Jaime, a man trained up in the business, tells O'Shay that she knows more people in the life than anyone else, that she was one of the first people to get Colombian cocaine into Europe, Australia, that she was a wizard at moving shipments and at being unnoticed. She handled heroin in fifty-kilo lots at times. And then she got caught in New York in the middle of a deal with half a kilo of heroin in her bra and went to prison.

O'Shay considers his business plan. He sees he has the marijuana, he has the cocaine, he has the airplanes, the trucks, the network. It is after midnight and he is home alone. He seldom sleeps early or easily, and he is entering his time of thinking. The light glows low in the back room off the kitchen where he sits on a black leather couch that curves and faces the stereo and large-screen television. Night sits right outside the two French doors. The walls are white, everywhere in the house the walls are white. Soft, pointless music purrs from a cable station. He cannot even hear the music but he thinks such sounds will calm him.

He sits with his drink in the semidarkness and goes over his plan again. Marijuana pays, especially the high-grade, hydroponic product, but it is bulky to move. Cocaine pays more and is more compact, but it is hell on the customers, who seldom last more than five years before they must back off. But heroin is forever. Heroin is compact. Heroin builds a base that he can count on. And heroin moves through the best hands, the most hardwired networks. If he gets into serious heroin, he will be dealing with serious people. He needs serious people in his life now.

Partly it is the money, but money is just a way to keep score. He has money because a Jewish businessman treated him like the son he never had and taught him how to invest, so money no longer tastes the way it once did. He can remember needing money and wanting money. But this feeling is long gone. A roll of crisp hundred-dollar bills has become dead weight.

He can get Mexican black-tar but he wants a purer and finer thing. He is tired of Mexican gang people, tired of simply making numbers, getting that weight and all of those dollars. He needs people with standards, a relief from the punks and morons he often deals with. And part of having standards is being willing to enforce standards. He will do it over heroin because he respects the drug's power to take away a person's spirit. He will do it with the Colombians because then he will know he can go up against the best.

He runs through the matter again. Colombian product will run more than 90 percent pure, and with such a product one can build a legion of users who will be faithful for life. He figures he can begin at a level of $10 million street value every two weeks and then build up from that. Initially, he can walk on the stuff ten or twenty times, cut it way down, until his customers' systems rise up to the new bar of quality and become at ease with it. The pieces will fit perfectly together: a Colombian network that is all business, no street bullshit, a group of professionals who deliver and expect payment. And a group of addicts, who, if they maintain minimal discipline with their needles, will last forever, live just as long as nonaddicts. And remain committed to his product, his quality, his constant and assured supply.

It all makes perfect sense. No one can question the logic of such a deal.

Besides, he knows he cannot fight the future. The Colombians are coming with their super-pure product, and once they reach the market they will win the consumers. The world of Mexican black-tar heroin will ebb and recede and vanish as customers learn to tolerate the high-grade stuff. So he will reach out to the Colombians and gain a beachhead in this future. He will strike first, he will meet the future at the portals.

He stops and stares into the black house, and a strange thought crystallizes in his mind: that he puts himself through all this hell because he has the ability to do it and in some way, if he does not, well, it will be a sin. Yes, that is it: it will be a sin because in doing these deals he learns more about humanity and about the greed and decency and the need to find something. He visits in these deals the farthest shore of love.

He stands, walks into the kitchen, runs his hands along the granite counters, eyes the stained glass in the breakfast nook, his fine, expensive range. Cooking helps, the chopping and flinging of things into a pan. Alone in his kitchen at midnight, the sizzle rising off the oil, he invents from random materials he finds in the refrigerator. He feels an ease at such moments, senses an opening up within himself that he seldom can get from business anymore. That is part of the problem: there is more pleasure in feeling a chef's knife cleanly chop an onion than there is in the routine flights of cocaine, the truckloads of marijuana, the suitcases of money.

The living room glows from streetlights seeping through the big stained-glass window that frames the house for him. In the corner is the baby grand piano, facing the fireplace is the white sofa. And on the walls the oil paintings he commissions, which are scattered throughout the house.

Almost all of these paintings are of waves hitting the shore, waves crashing and snarling, white foam shooting up, the water a green blue of anger, and the sea stretches into an infinity back from the shore, reaches out and beckons to another kingdom of energy and force. The light is always feeble in these works he commissions, no break in the sky where a yellow column of sun dances on the tumult, he cannot abide that. Just the dark and angry sea hurling itself against the rocks, the cliffs. The paintings calm Joey O'Shay.

He enters the hallway where his own paintings hang. A man in a very old style of baseball uniform sliding into base, the grass a flat, adamant green. Ducks landing on water at dusk surrounded by woods. A boy floating in a jon boat on a lazy body of water, a boy staring out of the frame as if his face were an unanswered question.

He moves on, back to his rear room with the music and the black couch, the French doors leading out to the patio. He has almost decided. He has figured out the details, walked himself through all the ups and downs. He prides himself on details, on never being surprised, on working it all out in his head. He has finished that part. He has drawn a plan in his head with moves, and with each move an exit, a kind of trapdoor through which he can disappear if things go awry.

Never want, he has taught himself. Never be eager. Never care if you do a deal. That is the power, and from that power comes the safety. But still, no matter how much you plan, no matter how many details you consider, there is always the possibility of a bad moment, and the need for the trapdoors.

He opens the French doors, steps out into the night. In back behind the garage is the servant's quarters he has converted into a weight room. But he does not head there. He stands under the vines lacing the ramada and inhales the soft night air. He sips his drink.

The city is so quiet at this hour he can almost hear the

knives sliding into flesh, and the bark of pistols where there are no trapdoors.

He shuffles forward out under the stars, walks the path to his garden. The okra, peppers, and tomatoes are coming along, the leaves brash. He runs his fingers through them, enjoys the rank scent.

He has had this thought which will not go away: that unless he does something, he will die. The thought comes from this sense within himself, a sense of death growing in his body like a plant, setting down roots and slowly sucking the life out of him.

He looks at his big forearms, can feel the power of his biceps, shrugs his shoulders laced with muscle. Yet his step is leaden, his arms heavy, and his mind seems filled with clouds.

If only he would go to the weight room and do his ninety-minute workout, he knows he would flood his cells with some kind of relief. But he cannot make the move. Nor can he run, run down the paths of the park, a razor stuffed in his sock, the trees huddling around the ponds, the city erased by the beating of his heart, footfalls creating almost a trance state in him, taking him to that place of clarity and ease, he cannot even do that, cannot simply change into shorts, put on some shoes, go out the door, and run.

He hunches his shoulders, feels his neck sink down into his body, the drink clasped by his fingers that should be wrapped around a weight instead. He is not depressed, that is too simple a word for how he feels. He is not frightened. He is not really worried, the constant tumbling of the deal in his head may be a preoccupation but it is not worry. He remembers when he was younger, hell, he was 192 pounds and cut—damn it, cut like a block of stone. Now he sees his gut spilling over his belt, stares at the drink in his hand, and feels life flowing out of his body.

He is very good at what he does and he knows it is the only thing he has ever been very good at.

But he does not believe in what he does, not any longer, not a bit.

Still, he cannot let it go because when he is doing it, he feels not simply alive but explained.

And so he will do something that means nothing to him because the doing of it means everything to him.

He sinks down into the couch and stares at the white wall faintly glowing in the black of the night. He knows he will be up until at least three, drink in hand, soft music playing, all those details marching by in his head, and yet at the same time he will be somewhere else, and in this somewhere else he will not be able to hide from this sense of emptiness in his life. That is what has happened: he can no longer deny the hole in his life.

He no longer has many moments of violence because his control is so sure that such moments seem precluded. The killing seems behind also. He is always ready to kill but his control seems to keep the slaughter beyond his reach.

Sometimes he envies the dead, recalls how they were ready, and because they were ready, they moved on. And left him behind, an unfinished thing. Just as he remembers those who have died, at moments that feel almost like some kind of communion, though he knows he cannot ever say such a thought out loud.

So he will make the deal. He will call that number he has obtained. He will talk to a man named Alvarez.

COSIMA CAN LICK THE DEAL, savor it. O'Shay will see. She is about to do the thing which brings her the most pleasure and she has found that deep pleasure comes not from love, not from fucking, not from food. Not from what others consider appetite. No, the deep pleasure comes from proving pleasure is not possible, that meaning is empty, that destruction comes like prayer.

The towers lean, the multitudes line up covered with sheets, their backs to her, the despised one stands before them with a beard and preaches. How she hates him as she works her painting brushstroke by brushstroke.

Yes, she can get what she needs from this deal.

Betrayal, ah, she can taste it.

O'Shay will now recognize her power.

And her devotion. She is not sure he will recognize the devotion, he is blinded she thinks to these finer things. Or maybe not, no, maybe not, because she can sense in him a kindred spirit. But he will not admit this kinship.

So she will show him.

And then it will slap him in the face.

———

O'SHAY LIKES WHAT HE HEARS about Alvarez. He appears to be a man of merit, a man who refuses to be a fool. Such a man can perform a deal correctly. And such a man can be bent, O'Shay knows, can be bent because such a man at the end of the day is always rational. Rational men are trapped inside their logic and their plans, and when these prove to be incorrect, the words stall in their mouths, they are paralyzed for a moment. He knows Alvarez is now in this place. He has had bad luck, lost a load, and owes half a million dollars to people who do not listen to explanations or tolerate excuses. He is anxious and that is when the power in a deal moves to the other side of the table. He is an exacting man, a person who pays attention to detail. And he is floundering, despite all his careful work.

He trains his carriers in Colombia. They are stuffed with roughage until their stomachs learn to tolerate new and cumbersome burdens. He takes his students to all-you-can-eat restaurants, then he moves them up to swallowing whole small baby carrots. He acquaints their systems with laxatives so that delivery seems normal and businesslike.

The students vary. Some can swallow and hold an entire kilo, some only half. But in Alvarez's calculations, they are all good earners. Each courier gets two to three thousand dollars, but each stomach carries a profit for him of tens of thousands and up. When they are finally ready, they swallow the condoms packed with heroin and take their seats on the airplanes.

Joey O'Shay has reached out to Alvarez through Cosima. O'Shay has leaned on Cosima, insulted her, told her he wants good Colombian connections and that he does not think she can pull it off. He has told her he will not go to fucking New York, the connections must come to his city and do business on his ground. Alvarez is the result of this pressure.

There is something missing in everyone, they wander and root and rob and steal and murder, they do deals, all in the hope of finding the missing thing and filling the big hole in themselves. There are two key elements, O'Shay knows, to every deal. Playing off this need of others to fill the emptiness inside them. And never letting this part of himself be put in play. Because once he wants something, once he reaches for it, well, then he knows he cannot have it.

But Cosima is his tool. She came to him with hunger in her dead eyes. She had begun her search early. Raised in a good family in Mexico, she ran off to a beach resort with drug guys and spent days fucking, sucking. The matter became a scandal with newspaper headlines about a kidnapping. This was a lie, she went willingly and gaily.

Her body developed very early, full breasts, tight waist, angelic smile, clear girlish eyes. When she thinks of certain moments, she has the habit of lightly licking her red lips. She worked her way up, first on her back, then on her feet. She became a courier, then a broker. She watched men executed in front of her. She fell once or twice but the sentences were light, the U.S. prisons easy.

By the time she entered O'Shay's web she was blooded, having moved 14,000 kilos of cocaine and a great deal of heroin. She had worked Mexico, the islands, Colombia, South Africa. She had been around murders in the city itself. And still she was empty. Like O'Shay, she took up painting, in part to be like him and in part to fix herself. Like O'Shay, she failed.

Her face is still smooth as a girl's as she slides into her forties. Her fingers gleam with three jeweled rings on each hand, gold bracelets on her arms, a gold necklace around her neck. Her breasts glow through the generous gap in her blouse, her small feet rest in her black boots. Her voice flows like honey, a voice that cannot be coarsened by any act or by any experience. Her voice seems to live a life independent of her own life.

Prison has given her some gifts. Her art began there, doing

tattoos on other women. The Hispanics asked for small crosses or hearts, the crosses or the names of children on the upper arm, hearts on a breast or, for the Mexicans from Durango, a scorpion on the ass. The whites wanted two red cherries splayed from a single stem with names in each cherry: the mark of lesbian lovers. Prison also gave her yet more connections.

She carries a book, O'Shay knows, and in the book are phone numbers, phones in Colombia, in the islands, in New York. When he meets with her and tells her what he wants, she lightly licks her lips, her innocent eyes underscored by black liner. As he talks, she twists a ring on her left hand, plays with it relentlessly. Or she flicks the pen in her hand, clicks it again and again as the point flashes in and out. Her body is always at ease.

Cosima lives as his shadow. He will be in an apartment settling a deal. The men are sitting around a coffee table, some having beer, some refusing to drink. The talk is soft, or at times the talk is harsh. Two of O'Shay's gunmen fetch the food and drink as required. And always, Cosima is moving in the background, never speaking, she knows better than to intrude on such crucial moments, but moving like a ghost behind the chairs and the couch as the men settle the deal.

He tells himself that she is motivated only by her endless greed. But he knows better even as he tells himself this lie. He knows she will make the calls and beckon the connection because it feeds the hole in her. She lives off betrayal. That is part of the attraction O'Shay feels. By having her in his organization, he keeps menace near. He watches her do deal after deal, rip off some people. And yet it does not touch this hole in her. The betrayal is never enough. She moves like a cat, the smile gleaming, and circles her prey.

Only one thing, he realizes, could ever satisfy her, could ever fix her. Doing him. He keeps this moment at bay, just as he keeps her near.

So he insults her, sends her off. And waits.

She does not fail. She tells Alvarez that Joey O'Shay is rich, established, has no record and a fleet of trucks. And hangars of small aircraft. This last is golden because the Colombians, like everyone else, know that small aircraft are beneath notice if they fly only within the United States. Also, Alvarez wishes to get into the online pornography business and O'Shay knows the business and the obscenity laws cold.

So Alvarez comes to the city from New York. Cosima meets him at the airport, they go back many years and have to catch up with each other's lives. She puts him in a nice hotel near the airport, and then they go out to dine. She tells him she wishes for him to meet O'Shay, that it will go easily since O'Shay speaks Spanish.

Cosima and Alvarez sit in the corner, light seeping through the windows, shadow washing their faces. The restaurant features Southwestern fare and squats in a row of such establishments off the big road. Alvarez and Cosima do not stand out, they never do. That is part of the business. There is nothing strange about this industry, it mimics other industries with contracts, deliveries, and payments. That is what both assures O'Shay and at times bedevils him. He finds it very difficult to separate the realms of approved behavior and criminal behavior. They seem twins to him.

Alvarez dresses well, his body compact at about five six and 145 pounds, his hair straight, his skin black, his manners excellent. He speaks crisply and softly and yet without hurry. He avoids business at the table.

When O'Shay enters the restaurant, Alvarez stands and shakes his hand. He thanks O'Shay for meeting him. His Spanish is Colombian but O'Shay soon learns that his English is excellent and they shift tongues. He notices Alvarez's nails are manicured and the hands look like a bitch's. He sees sincerity in the eyes. He swallows Alvarez whole—color, smell, sound, features.

O'Shay senses that Alvarez expected a softer man and is surprised by his bulk. He can tell in a handshake that Alvarez has never worked with his hands.

They talk about the weather, they talk about food. Alvarez slowly sips his wine, O'Shay drinks coffee.

Alvarez wishes to pay for lunch, but O'Shay says, "No, no, you came here for us, you are my guest."

The moment comes for real talk and so the two men leave the restaurant and get in the front seat of O'Shay's car. Cosima, as broker and underling, gets in the back.

Alvarez says, "What is it you want?"

O'Shay replies, "Look, I'm going to be honest with you, I'm too goddamn old to fuck around and you came too far. We've got a heroin market here, a real good thing going. We need better heroin, the stuff I'm getting from the Mexicans is declining."

Alvarez says nothing, he sits staring forward, his black skin cool, his manicured fingers at rest on his lap.

O'Shay continues without skipping a beat, "What I am worried about with your connections is that the heroin will be so much more powerful than what my customers are used to. It will have to be cut and they will have to be educated to it."

"Yes," Alvarez says, "it is very powerful, it is deadly and will have to be cut. I've brought two samples, an ounce each."

He reaches over and hands them to Cosima in the backseat. O'Shay takes note of the etiquette, pleased that Alvarez does not insult him by handing him the heroin.

The meeting is over.

Later, they meet again in an apartment. They sit with a bowl of chips.

O'Shay says, "Look, you know what I can come up with. I'm

not in a hurry. What I have here is silver, the cocaine, and what you have is platinum, the heroin. I know you can make a lot of money. I have the trucks to get your coke to New York."

Alvarez says, "I would like to get two hundred kilos at a time."

"Well, we can do that. No one wants to fuck with a truck-load of rocks, and to get to what we would have in there, they'd have to unload fifty thousand pounds of fucking rocks. And even then they couldn't find it. We're interested in getting a good source of heroin because we think there will soon be Russians with southwest Asian heroin here that our stuff will not stand up to."

Alvarez interjects, "I have access to China white heroin. But not a regular source."

O'Shay continues, "I don't really care so long as I get powerful stuff."

The meeting is over, the chips barely nibbled. The voices have never risen above a monotone as flat as the city itself.

Cosima takes Alvarez to the airport.

O'Shay feels possibility rise in his mind.

HE DRESSES IN BLACK with lizard-skin boots, the sports coat his only light note. He keeps turning the deal over in his mind. So many details to master. The gun rests comfortably at his waist. Always chambered, no safety. Joey O'Shay is ready.

He is looking at a deal that would gross several hundred million dollars a year off this one connection. And then it would grow. On paper, the deal dazzles him because he thinks no one else could pull it off. There are people moving in Colombia because of him. There are people moving in New York because of him. There is his own kingdom right here and people in his city are also moving because of him. His hand holds the strings and when he moves that hand, others dance. He likes this feeling.

But today at lunch he cracked a crown and had to call a dentist. Also, his drinking is growing, and he does not approve of this. He is now drinking on weekdays. He is breaking his own rules.

He considers these matters, takes their measure, and then dismisses them. He has never failed. He will make the deal.

He has always promised himself that he would get out at fifty. Time is growing short. This will be his way out.

BOBBIE LAUGHS AT THE IDEA of O'Shay quitting. She snorts, "He's an adrenalin junkie. He'll rest up, he'll be back. It's in his blood. He's always pushing the envelope. He's got that big boat and yet he's never on it. Painting? His painting isn't going to keep him happy. Fishing isn't going to keep him happy. Where does he go? What does he do? You take that away from him and he has to be a normal human being. He can't. So where does he go?"

COSIMA WANTS IN, she always wants in. She will do anything to be in Joey's deals. She has made the initial contact but since then she has been left in limbo.

Bobbie does not trust Cosima but this feeling she has is not jealousy, no, it is simply awareness, a sense of knowing that Cosima must have blood in order to feed herself and that she will, without doubt, get Joey killed. Bobbie is sure of this fact. She is more than sure, she is certain that this is the reason Joey keeps Cosima around.

Joey rolls down the freeway at night, his phone rings.

He listens and says, "Puta, why are you fucking bothering me with this, with a couple of fucking keys of heroin down south."

He snaps the phone off but he thinks, The bitch is working, Cosima is always worming her way in.

———

O'SHAY SITS EATING off to one side of a café, a platter of mixed seafood in front of him, beers coming one two three four like soldiers to battle. No one is near him, his back is to a wall, and he cannot easily be seen from any window. This seating has happened as if by accident but still, it is his habit of seating himself.

He clings to memory and yet erases the past and does this as one seamless act. The past cannot interfere with the present, but the present cannot exist for him without his past. So O'Shay is always of the moment and yet not present in the moment. He listens, speaks, acts, and when he acts it is very fast, almost fluid. But he is not there, he is in some other place. This place exists without a presence on maps. Always, should he look up, the stars are overhead and there is without exception scent in the air. He is floating and hears the cries of night birds, the soft roar of midnight insects, the faint stirrings of fish, the lonely songs of falling stars. And voices, always voices.

HE IS IN A ROOM WITH TWO MEN. They are talking business. The door is open to a hall.

One man says, "Billy, Billy, why you no show me any respect? You think I afraid of you? I not afraid. The *federales* not able to make me afraid. See, look at my scars, I pull up my shirt, look what they do to me. I not afraid. So Billy, Billy, why you no show me respect?"

The man called Billy is silent. He listens with a blank face to this recital.

Then suddenly he upends the table separating him from the man showing the scars. He knocks the man to the floor, and begins kicking and stomping him.

O'Shay gets up and silently closes the door.

The man on the floor says, "Billy, Billy," as the heavy boots smash his body.

O'Shay does nothing.

It is the business.

JOEY O'SHAY IS DRIVING the city at night, his time of comfort. He can barely believe in anything before dusk. Even in his paintings it is almost always night or soon to be night. He thought the paintings would help him. He taught himself everything, worked at it like a slave. But the paintings failed him.

He is talking as he stares straight over the wheel, gliding easily from lane to lane as he threads through traffic. He can vanish into a monologue at such moments. He is always well spoken, the sentences measured, planed, and smoothed, laid down one after another like carpenter's work. When he makes eye contact, he lies with ease. But driving, especially at night, brings the truth closer to his lips.

He is a natural liar. He knows most people cannot manage their lies, at least not very well. In part, there is the manner of tells, those signals they give off when lying, the darting eyes, the sweat, and should they master those giveaways, still they persist in the false tones, the claims of sincerity, and, of course, the ultimate weakness in such matters, the deep need to be believed even when lying. He feels none of this and gives no sign when he lies. He has this edge because he knows the other liars are really shopping for a moment of truth, seeking like the lovelorn that moment when they can tell someone the truth and still be accepted, still be forgiven. Also, they wish to be respected, to be known, in a way to be famous. They want a reputation, the good table in the cabaret, the special girl in the clubs. They want to be somebody and cannot conceive of being somebody unless others know who they are.

He shares none of this, he despises such appetites as weakness.

But at the mention of this one thing, this small thing that

cannot be said, something about a boy, he freezes, sputters, almost shatters like glass. He falls into silence, his eyes lock on to the white lines.

Then, after a few moments, he pretends the matter never came up and he goes back to the deal and the details of the deal and the comfort of those details. He is his work or he is nothing. This he knows.

O'SHAY KNOWS, from novels and preachers and politicians, that the accepted key to life lies in what underpins a life, some timeless truths that guide a person and infuse all actions with meaning and value. Life becomes a fulfillment of these truths and living is simply acting them out as yet one more player in an endless pageant of the faith. Marry right, raise the kids, pay taxes, work hard, set a little by. And be safe. The custom is to mind one's own business, don't cause trouble, and obey the law.

O'Shay cannot do these things nor believe in the doing of them.

The night blurs past the window, hungers arise and scrape the throat, pound the heart like a drum, the pedal is to the floor, everything feels right and nothing is permitted and all is allowed and suddenly another country appears, clear and pure and fresh as springwater dipped from the flow on that first day of summer. And after that taste, the unforgettable taste, nothing is the same, not after things go bad, not after knowledge comes and pollutes the water, not when the door on the cage slams shut, not ever. Regrets are a piety, but appetite is the engine.

Joey O'Shay enters this country. He will never be able to state exactly when. He will, perhaps, in an idle moment speculate, imagine some moment, some decisive event, some turning point. Then he will push this thought aside. Almost everyone who enters this country does. That's the rub: there is no turning point, there is hardly a memory of a turn, it all seems like a straight path, and only later does the belief come that says there was some turning point.

But for reasons personal and not so personal some go down that path.

They can never quite say why.

But they always wonder why everyone does not take this trail.

Joey O'Shay does not hesitate.

He tries to sort it out. Not apologize for it, just sort it out. He can kill. He can maim. He can dominate. He has the gifts. He can act without hesitation, move with fluid grace with an ice pick, a pistol, or a fist. Sometimes with a straight razor.

He has trained and trained to perfect himself. He has studied. But he knows, his real assets are a gift and he must not fail these gifts. That would be a sin.

He tries, he is always trying in secret, nobody watching but himself, trying to make sense of it, of this ability and appetite and lack of hesitation.

...to abuse the gift is a sin that can only be forgiven by God the father. To eternally deny it...to take advantage for our own advantage is...to have no redemption...

Here he pauses, the word *redemption* attracts him and yet seems a stretch.

...there you have an escort in your decision and end. From the shadows it will escort us...in all its ugliness...and agonizing sadness...to observe our wasted lives forever alone...

The thing, that hulk, O'Shay cannot abide that notice is not taken of this fact. For him evil is a word that must be rescued from the fat pastors braying on Sunday morning, must be taken back and returned to the streets where it roams and lives and thrives.

...it slides by sometimes, a feeling...a smell...a rustle in the alley...an uneasy awareness. Sometimes it peers or makes its grotesque self seen or heard. To remind us...of our kind's ancient blessing and task...carried in our hearts and blood...our earthly duty...from God...

HE DOES NOT HAVE TO DO what he does and that catches the attention of Joey O'Shay, who is at that moment a beginner in the drug business. Angel Sonrisa comes from a moneyed family, has the schooling and yet he is out here living the life, the streets and bars and tired cafés. He becomes the rabbi.

Christ, O'Shay thinks, this guy's in his mid-thirties, how in hell does he still do it? And why is he out here? O'Shay, all muscle and strut, is in his twenties. He's mastered the rudiments of his business, learned the ins and outs of various products, how to test them, where to carry his weapons, what to watch when he is in the room. Sonrisa is a master and O'Shay knows it and so despite the decade separating them and despite the difference in their backgrounds, he becomes the eager pupil.

The lessons are very simple at first but always very hard. Patience, be patient, Sonrisa teaches. Never hurry, no one who is real in this life hurries. You can never be hurt by the deal you walk away from, you can never be wounded because you sensed something and did not show up for the meeting and did not go into that room.

For O'Shay, this simple matter of pacing is hard. He likes to work alone because then everything happens at the right pace, his pace. The world in general operates too slowly for him. He is not designed to sit in a room and listen to a lecture and for that reason, he is largely self-educated, both by books and by the street.

But Sonrisa insists on this patience matter. In part, he explains, for personal safety. Anyone who makes fast moves in the business becomes a victim of his own desires.

Desire is always read as weakness. They smell your eagerness and they think something is wrong, a serious player would never be so eager. They will back off and shun you. They will not be exactly certain why they do this, but that hardly matters. Certainty is rare in this business, everything is about odds, and about engineering the odds to the most favorable possible outcome.

Those who seek absolute certainty, who demand total security, are incapable of the business and of the enormous returns made possible by the sheer risk of the business. You caress the odds, try to tilt them in your favor, but still you are at their mercy. And so in this place where certainty does not exist, where contracts mean nothing unless there is one on your life, where you do not get to call a lawyer to straighten out a business deal that has gone sour, in this place you must risk. And seeing desire in the eyes of someone you are dealing with raises this risk, because it is not a sensible emotional state and people that unbalanced could be dangerous.

O'Shay learns. He does not hurry, he never pushes.

He also learns that eagerness can get a person killed, that the hunger on the face becomes the trigger for a rip-off, and in such cases someone will certainly try to take his product or his money and perhaps also his life. He becomes calm, not serene but calm, he becomes patient.

He knows Sonrisa is the man to study, that anyone who has survived these streets must know and that the things he knows will not be found in books or even on paper. He knows he must learn by listening and watching, that it is a set, almost an endless series, of little moves, small gestures, a dictionary of body language, of something in the other person's eyes, that what he is learning is a life that must be led twenty-four hours a day and seven days a week.

Be yourself, Sonrisa advises. Never stance, never lie about yourself. Just be yourself. Don't pretend to be more than you are, people will smell the fraud coming off you. Don't be tough, just be natural. If you are reading some cookbook, then you are the guy reading a cookbook. Don't pretend to be reading something else. Because the trick in this world of deals and frauds and rip-offs is this: Be trustworthy. Be natural. Don't talk about yourself, don't tell others anything, absolutely nothing, that they don't need to know for the deal. But given this curtain of privacy, still be natural and open. If you do not drink, then refuse the drink. If you don't like titty bars, say so. If you don't know shit about bikes, admit it. And for God's sake, don't pretend to be some tough guy because in the room someone eventually will test your claim.

So O'Shay is on the streets, in the rooms, he's making bones, doing deals, and now he has found the perfect master. He begins to live the life more deeply, he absorbs a hundred little details at a gulp.

They'd had a running argument over weapons. They both packed guns, but Sonrisa had a switchblade as backup and O'Shay preferred an ice pick. The trouble with a switchblade, he'd tell Sonrisa, is that you have to flip it open. The ice pick is ready for work as soon as you bring it up. O'Shay knew he could put his ice pick in a man's eyes before the guy could pull his gun, much less fire it. Once O'Shay saw a dead guy who'd taken fourteen in the head from an ice pick and the work had been done so well there was almost no blood, just a little dripping from his mouth.

He almost envied Sonrisa's good nature. He seemed always

to be getting married and then divorced but this never broke his stride, never took the edge off his attention to the business.

Only one thing Sonrisa never explains: why they are doing it at all. O'Shay has a small set of reasons if pressed but they hardly convince even him. Sonrisa has no reasons since he does not need money. But, still, he is out there.

HE FEELS SPLIT. The city spreads out to the horizon—small houses, hot streets, fumes. The city invites collisions. The piney woods give way to the prairie, the warriors of the plains meet the lances of the woods. Cotton dribbles into wheat, hogs give way to cattle. The plow stops and is crushed flat by the mill. There is a history but no one attends to it because the city rises as a crossroad that erases each moment as it lunges toward the next. Here the north slaps against the south, the east is scraped raw by the claws of the west.

He climbs a fence and drops into a yard full of beasts. Suddenly, a dinosaur charges. As the monster closes on him, two hands lift him up. He is four and it will be some time before he realizes the dinosaur was a goose.

The city first came into being because of the river and because of the roads, but for him the city is always creeks, small secret fingers of musk and mud and slow water wandering unloved through the warrens of people. He moves down the bank through the thickets that thrive in the city, pecans, oaks, mesquite, a riot of trees. The streets vanish from sight and mind, fish glide in the waters. A thrum of insects clogs the air.

At times, especially dark moments like holidays, he remembers the taste of those miniature days that continue to come unbeckoned and loom so large. When he returns to childhood his family seems absent, his father and mother hardly present, his brother out of sight. The only thing left in this landscape is a creek, tucked away like an intestine inside the city.

The creek barely merits that title, a quarter mile of open storm drain spilling its flow from a big cement pipe into a deep and rocky pool. The water came from the north and after leaving the pool enters yet another big drain and flees under a boulevard. This is the ground of his gang and where it ends the world of another gang begins.

The creek is everything for O'Shay, a dab of water four to ten feet wide, sluggish and rank, maybe a foot to four feet deep, green, and buried by trees, vines, mosses.

...some of the trees must have been ancient as they were enormous in height and girth. The tallest cottonwood tree and sycamore tree I have ever seen loomed in the middle area of the woods.

The woods harbored a vast array of wildlife for its size. There were cottontail rabbits, fox squirrels, bats, and field mice. There was a great variety of birds that included small owls, woodpeckers, dove, and all variety of songbird. The creek itself harbored numerous species of frog including tree frogs, large leopard frogs, and the occasional bullfrog. It also had small minnows, tadpoles, salamanders, and crawfish that got up to six inches long. The area was abounding in reptiles that include numerous types of snakes, big striped racerunner lizards, horn toads, skink lizards, green chameleons, with big grey wood lizards (in the trees). All manner of insects buzzed about including various species of squealing, singing, clicking cicadas, katydids, crickets, and grasshoppers. A million varieties of moth and butterflies hung around the flowers, mulberry trees, and shrubbery...

A dark form, the hump tiled, the color wandering from gray to brown, the movement effortless. It is dusk, he stands on the bank, the huge form moves past.

When he is eight, he sees the same form in a city aquarium,

Macroclemys temminickii, the alligator snapping turtle. The record specimen ran two and half feet in length. One caught in the wild weighed in at 315 pounds. No one knows how long the beast can persist, fifty years, perhaps a century. They sleep on the muddy bottoms during the day, rise up at nightfall for the hunt.

For decades, he will return to that moment, the quality of the dying light, the grace of the beast's movement, the tension and yet peace that filled his body as he stood and witnessed.

Joey O'Shay will carry the creek inside him like a suitcase and sometimes he will open that suitcase and savor its contents but he will never merge those contents with his work. The moment will remain distinct, almost jewellike, and have the smooth facets of a jewel.

He goes to this memory inside of himself to touch some kind of ground but he shelters this memory from his life.

O'Shay hooks up with other boys, belongs to a pack that roams the streets and thickets and creek bottoms. He learns about grenades, builds one from old junk and shotgun shells, detonates a garbage can and alarms households.

A boy betrays O'Shay's pack. They kidnap him, take him into the woods, strip and gag him, pummel him for good measure. They tie him to a tree and leave him. O'Shay lies in his bed that night and tries to sleep but he cannot get the kid out of his mind. He slips out his bedroom window, moves through the thickets in the darkness, finds the boy and releases him. He makes him promise not to rat him out. The boy agrees, then rats him out.

A harelip is in his school and the class bully ridicules him constantly as O'Shay watches and does nothing. Finally the fight comes and the harelip pounds the bully to a pulp. O'Shay feels a shame that will never leave him. He has choked and gone along with his gang and he knows it.

O'Shay can remember catching alligator snappers on the edge of the creek, recall the stench of them, taste the air, hear the din of the insects. He can recover all this. Some things stand out and there is no need to ask why.

...then I saw it...a huge turtle...glide so softly underwater from the deep undercut banks...then glide down that narrowly beautiful chasm of a creek...and into and under the opposite side undercut bank. It was a big snapper...two feet and thick. I was astonished. I was blessed. I couldn't believe it was so huge in this tiny little creek...this tiny little world.
...perhaps he is still there...silently majestic...

The alligator snapping turtle goes largely unnoticed, a thing of the night hardly heeded when present or missed when gone. All over its range it is receding as men drain swamps and dam rivers, as men harvest the turtles and take them home and eat them. O'Shay catches them accidentally while fishing for catfish in the creeks and elbows of rivers. He carries snippers to cut the hooks from their mouths, careful not to lose a finger, and always lets them go. Even the young he discovers, tiny three-inch hatchlings, try to bite him. Other turtles just lie there patiently and wait for him to cut out the hook. But the alligators act with fury.

They are the only reptile in the world to lure their prey. They lie on the bottom, in the mud, open their mouths, the tongue waggles and looks like a worm, the fish glides near. The jaws snap. They do this again and again, in the night, in silence, hunting out of the mind and sight of the city. Slowly, as O'Shay lives, the turtle's numbers dwindle and teeter toward some state of being called endangered. By the time O'Shay is a man roaming the streets of his city, the alligator snapper is vanishing, in some places gone from 90 percent of its former range.

HE DOES A DEAL and the guy has this self-made video in his apartment, the kind of thing made for private enjoyment. It becomes a memory O'Shay cannot shake.

The movie begins this way: She looks up through the clear plastic bag over her head and makes soft animal noises as he slowly suffocates her. Her skin glows a rich brown, the hair black and uncombed. She wears at the moment a blouse, shorts and she is barefoot. Her legs draw the eye as they writhe and at times have tremors almost mimicking the onset of rigor.

The guy seems more at ease. His shoulders broad and powerful, a casual sleeveless shirt spreading across his back. His hair is also black, his skin brown, and his face not visible.

He will choke her for a spell, that is when the soft and yet not quite human sounds arise, and she will struggle, not actually fight him, that seems an unthinkable thing, but struggle, her body writhing, her legs thrashing, her hips grinding into the floor but none of these movements suggest ecstasy.

The bag is translucent.

At first, everything looks threatening and yet playful, the dance of suffocation to evoke sexual excitement, the sport of lovers after they have caressed the surface of their fantasies. Something like that. But this impression passes. He holds the bag too long, she fights too hard. The sport drains out of the scene.

Every few moments, he will lift the bag and then her face suddenly emerges as if from a grave. Her teeth uneven, eyes wide open and wild with fear, and the face young, a sixteen-year-old face.

She will say, "Why are you doing this?"

He will answer nothing in reply.

She will say, "I don't like this."

His silence falls on her slender girl's body like a tire iron.

Sometimes he makes the faintest gesture with his shoulder, a slight dip and very small turning of his head, and this almost toss of the head seems clearly directed to whomever is operating the video camera. There is almost a gleam from the man, and this is impossible to explain because his face is never on camera, but still the gleam is unmistakable. He is looking back at the person running the camera and his body says, See what I'm doing?

And the camera has an operator, this is certain. The camera moves in and out, the camera tracks. The camera has an intelligence spellbound by what is happening as the girl suffocates under the transparent bag.

That girl is looking up, her eyes pleading.

The man's back straightens, his large hands come down again.

The choking resumes.

BOBBIE PULLS O'SHAY aside into her office at the hotel.

"You've got to let me into this deal," she says. "Let me do this. I need it to stay alive."

O'Shay is not used to having her ask. For her to ask for something is like begging. And she does not beg. But now she is asking.

He nods. Things are not right. The crown broke in his mouth. The deal with the Colombians keeps getting more complicated. And now Bobbie is asking for a piece of it.

For a moment, he feels rushed. Then he steps on that moment as if it were a bug, crushes it beneath his heel. Ah, he is better now.

"We'll see," he tells her.

HE LOOKS SMALL to O'Shay, too damn small, and he thinks the squirrel must be suffering from hunger or something. The animal huddles at the French doors and O'Shay gets off his black couch, goes into the kitchen for some cereal, and then tosses it out to him. The small squirrel gobbles the food.

One day he takes a basket of wash out, leaves the doors open, and when he comes in he sights a rat on his coffee table, damn near jumps from the surprise. But it is the squirrel sizing up the place.

He comes to call the tiny squirrel Geezo and will say things like, "Goddamn it, Geezo." He knows a Marielito Cuban by that name who is always sneaking around.

Geezo gets more confident and will wander into the kitchen alone, hop onto the counters and examine things.

Other squirrels show up, all bigger than Geezo. One especially seems to be the patriarch, and O'Shay calls this huge squirrel Tubs. He also seduces O'Shay and eats from his hand.

O'Shay feels comforted by the two squirrels. Birds and other creatures have always come to him and never fear him. Domestic animals tend to hate him.

He can feel gentle when he is with wild animals.

He is in a nightclub with a woman who is wearing a thin-strapped dress. A man comes over, and the man comments on her breasts and touches her strap and lifts it. O'Shay grabs the man's head and slams it into the tabletop three times. The nose goes flat with the first smack.

The man crumples to the floor.

Afterward, the woman asks O'Shay, "What would you have done if you'd killed him?"

O'Shay says, "Body into a Dumpster, no identification."

She looks puzzled and asks, "What?"

"You heard me."

HE IS FOUR YEARS OLD and the big backyard rolls down to the mulberry tree. His bedroom is small and spare on the back of the house and the glass door opens onto the endless yard. He is sick with measles, his fever high, and strange images stalk his mind. His mother is crying because the doctor has told her that if he does not get better, then he must go to the hospital.

He vomits, he is dehydrated, and he floats in and out of consciousness.

He thinks as he lies there that he hears his mother say that if he does not get better this night, they will have to take him to the hospital. He thinks he hears his mother say that he is dying.

A face leans over, the child smells the creosote coming off the face and knows it is his father who climbs telephone poles on his job.

The face asks, "Is there anything you can keep down?"

"Watermelon," the child whispers.

Then the child blacks out. He can sense his parents entering and leaving the room throughout the night.

But mainly, the child goes to the zone and in the zone an immense black cloud floats down, smothers him on the bed, and the child knows in his bones the cloud is death. He feels the weight. He is being crushed.

He pushes with his feet but he is too weak and the cloud does not move. He is above the bed and sees himself barely breathing as the cloud descends.

Outside the door, he sees this thing, a dog? A monkey? He cannot tell. He is not afraid of it and then suddenly it is by his bed and it is enormous. He stares into the face of the thing and it has the look of a dog and yet at the same time the look of a snake. Unblinking eyes that burn.

The child realizes he is going to die. The thing hovers, leaning into the child's face, and the boy smells it and it smells like something the child had met in deep, dark, dank places. A smell the child never forgets.

That night the fever breaks. In the dawn, the child's father appears with a watermelon and the boy begins to mend. He puts the memory of that night away, buries it out of view of his life. But the thing returns at odd moments of his childhood, will be there right by him. And then leave yet again.

But when he becomes a man named Joey O'Shay, he catches whiffs of that odor, the stench that came off the thing in his room. And he turns then to look, knowing it has returned and is very near.

HE CANNOT ABIDE WEAKNESS in himself. He's been moving furniture since he was a kid. This work built muscle, slowly turned the tendons into strong cords, gave power to the hands, and laid down calluses that shout hard labor to everyone he meets. So he'd had the base.

He'd done his time in the army, too, and goddamn, what they fed him made him lose weight. But he learned extreme self-discipline and survival skills.

So he set up a routine to shock his body into being, a routine that became his lifeline.

He goes to the garage, hits every body part with each workout, none of this resting shit for him. If it is cold, he wears lots of clothes, and if it is hot, dangerously hot, he works out anyway. He warms up by stretching real good, then slams down seventy-five push-ups and one hundred sit-ups. Then, without any pause, he hits his reps.

Sets of curls.

Three backward curls.

Three dumbbell curls.

Three sets of military presses.

Three sets of behind-the-back military presses.

Three sets of bench presses, sometimes with massive weight on the bar.

He never tries to set records, he tries to punish himself. And so after a while, he is curling big weight in endless reps in each set.

He finds his rhythm, his pace set by some internal signal, and arrives at a place where he works out for ninety minutes with only sixty seconds between sets. He never really rests, he never quite gets his breath, and with this barrage on his cells his body comes back and he can feel the strength flowing through him.

Twice a week in the beginning, after lifting weights, he'd go to the track and run sprints until he had nothing left inside him. And then he'd run some more sprints. Now he runs every other day, on his rest days from the weights. He has outlined a course that reels out for nine and a half miles through the city, into the parks, and along the languid creeks recalling his childhood.

He is not a bodybuilder. He is not about to enter contests and see who can lift the most. He is not interested in winning little ribbons at jogging events. He is sculpting a tool, one that never tires, one with broad shoulders that anchor the battering ram of his head. Thick stubby fingers ready to choke a throat or gouge an eye. Big forearms that rest on a tabletop, the kind of arms that whisper to the other person that he can tear their head off.

He knows the fear people have of being touched, how contrary to all logic they fear a blade more than a bullet and that what they fear is the intimacy of the knife or razor, the feel of the killer's warm breath on their body. He knows this fear is a tool if one is willing to use it.

He goes to the garage, follows a regimen of his own design, and pounds his body for ninety minutes.

He is sleeping only about three hours a night now. His real rest comes in the workouts, in the serenity of the reps, the sweat pouring off him, the blood rushing to the muscles, the breath not labored, the air devoured as his appetite for oxygen keeps expanding along with his body.

He tries to keep his drinking to the weekends. He uses no drugs, he sees the eyes of those who shoot up to gain strength and he does not want to have those eyes.

Years of moving furniture, of having just enough money for a twelve-pack of beer, cheap beer that is on sale. Odd jobs here and there, Christ, hauling silage at night to get a little more money.

There are these roles, these careers, these classes. There are these assumptions about education and skills and credentials. These mean nothing to him. The people created by the schools look like prey to him. The people in the suits and fine cars, they looks like deer in his headlights.

They refuse to know what can be known and so they are fated to go down. And he is building the hands and shoulders and heart and lungs and arms to rip their lives to shreds.

When he runs those nine and a half miles, he always goes alone. He cannot stand to do things at the pace of others. With time, he is clocking seven-minute-and-fifteen-second miles and hauling a hundred and ninety pounds of muscle. He likes to run late at night when no one else is out there, with a straight razor strapped to his body.

He runs, once for fourteen miles, but has no interest in completing a marathon.

He lifts but lifts alone.

He is young, he is nothing, he has no name on the streets, he simply races through them in the darkness with a straight razor waiting for a face to match its edge.

BOBBIE COMES OUT OF THE HOTEL and sits on the bench. The air sags with humidity—the city edges onto the lip of summer—and tongues of moisture promise hot and sleepless nights. A steady rumble comes off the freeway facing the hotel and the bums forage. Behind the hotel, a neighboring building has shut down its reflecting pool to stop winos from bathing.

She lights a long thin cigarette, inhales greedily, but the release is slow, very slow, and elegant, a thin trail of smoke pouring forth from her sensuous mouth. A bum stalks the edge of the parking lot near the grackles feeding on the strip of lawn banking the asphalt.

She grimaces at the sight.

For months, he has been wandering into this Colombian thing, branching out to new networks, sources, and products. O'Shay, she thinks, might as well be a fucking junkie, he can't seem to back away from a deal. He never has enough, he keeps reaching out. Well, no matter, she thinks. Her hotel will soon be clogged with Colombians and Dominicans.

She met O'Shay years ago, when he was at the beginning of his rise, renting an apartment in this huge complex she managed, and he could tell at a glance what she was, that good-time girl who floated on the edge of the local syndicate, that knowing player who had the straight job but the life of the other world. But she couldn't peg him until he told her. There he was, right under her nose, paying rent to her, operating out of that apartment and making hundreds and hundreds of deals there. Down and dirty deals, not like now, but still deals that mattered to him because he was determined to rise, and Bobbie with her quick eyes and fast tongue never had a clue as to what he was really up to and where he was really headed. And when she finally found out, well, she decided to come along for the ride, to float her life in this new stream of energy she'd encountered to her surprise.

Now she has a role that others cannot easily determine. O'Shay writes the script for the little play he is going to produce, and Bobbie becomes part stage manager and part bit actress. She provides lodging where dealers can be watched and yet be undisturbed. She does walk-ons as a woman who knows drugs and the world of drugs. She is the fixer who finds the perfect limo with the driver who neither asks questions nor misbehaves. She is the gal pal to the women who show up, women with high-toned tastes who must shop and need an escort who knows where to go.

The best shopping is only about two miles away in the mon-eyed island of the city, a place where O'Shay sometimes sits and watches the rich and imagines their vices and flaws and little tiny appetites that enable someone else to pull them down and fucking drown them in their own filth. The women always fascinate him with faces ten years younger than their bodies and bodies years younger than the bones holding up their flesh, women with expensive quiet clothing, small jewels that announce their station and eyes that always, always broadcast their fear of time and sagging and losing everything, even though they would be hard-pressed to explain just what this everything might be.

Bobbie thrives in the normal world and makes a good living.

She is not an informant, she works for no one but herself. And of course, Joey O'Shay. She takes no money for helping him. And this last matter she does not explain to herself. She simply knows she needs what he provides, needs the electric charge in the air. So he comes to Bobbie and she always says yes, says it without hesitation.

COSIMA PICKS UP THE BRUSH and the color red comes to her. She hesitates, then applies the pigment to the canvas. She knows O'Shay does this and so it is a way in. She has the deal working down on the coast but when she called about it, he called her whore and cut the conversation short. O'Shay needs her but he fights this need and she will make him want her again. Yes, she will. She is his snitch, she is paid, it's a job. But it isn't. It is a need. Every time O'Shay cuts her loose, foists her off on other agents in other cities, he fails to get rid of her. She comes back and he is happy to have her back though he cannot admit this out loud.

They share something unspoken: they are both natural predators.

Now her connections can take him into the deep world of heroin. But she knows, once he enters he will try to wall her off, to go on without her. He will try to limit her participation. And this she will not permit. Because . . . he belongs to her.

Two red rectangles rise on the canvas, red towers scratching the sky.

They are now red, blood-red, but the sky is heavy with storm or trouble. And above the towers, a bolt of lightning.

Yes, that is it, she thinks.

She hates them and what they did to the towers.

Once when she was there, she saw the towers off the wing of her plane at landing. Then she forgot about them, she was immersed in business.

But now the towers are back. She will never forgive them for what they have done.

That is why the towers are not simply red, they are blood-red.

TIME FREEZES AND SITS THERE as a bar of nothing. Or time gets caught in an eddy and twirls around and around and around, going nowhere. Or time...simply stops and becomes the stale air of a motel room, the blank eye of the curtains pulled shut, the phone that does not ring, not this hour, not the next, not for days as the wait continues. The television is on but the sound is off and images flutter, the curtains stay shut, the knock at the door means a pizza delivery, and the wait continues.

Within the heart of the business is not money or violence or death or drugs. Within the heart is nothing, the absolute nothingness of the wait that forms the core of the business. The deal is in motion, the players are moving, but in the city where O'Shay waits, time is frozen as the wait deepens and swallows everything, swallows O'Shay, Bobbie, everyone touching the deal.

Phones ring, tires turn, quick meetings occur, days roll by, and all of this is nothingness until people in other places and nations emerge and move.

No one can be in this business without tasting the wait. Boredom floors every day and night, the body is poised, alert, ready for action, and the action is to endure boredom until that key call comes, that key meeting occurs, the shipment arrives, the crisp bills riffle in the hand. This is O'Shay's moment, the wait, because in many ways it does not exist for him. He is never bored, never chained to the rhythms of others. He is never really waiting. He plots, thinks, notices, reviews. He does his own kind of patrols, visits snitches, hears tales of the street, stockpiles information. Waiters working in the good clubs tell him stories. Whores walking the street come over to his car when he beckons and then they lean in and report. He has people scattered throughout the court system, a small army in the probation world, other legions that handle sentencing. O'Shay wants to know everything and control everything.

The phone starts ringing, bringing his various identities into play. A rich man needs some bodyguards and O'Shay provides him with off-duty cops. A drug deal is advancing as O'Shay purrs into the phone, the man in this city to see for major action. A dealer he once busted calls with some information, the kind of tip every felon hopes will one day prove a get-out-of-jail card. Some case is stalled, one that has nothing to do with O'Shay, but he is asked to do the interrogation and he will. And usually the person will talk because O'Shay seems able to will confessions. Then he is off, rolling down the streets, drinking in faces and actions, pulling over for a few quick words and the money is handed to some snitch. He'll make that call to straighten out a problem for a junkie, he'll have coffee with dealers, meet with people in other agencies, go by the office. The day wanders without waste, a series of small tasks that can go well into the night and when it finally ends O'Shay knows more, has granted favors and has more people in his debt, has fixed things for people of various classes. Has learned some new quirk about human nature, felt his way into some intriguing new network. He is not making a case, he is building the foundation for doing any case he might desire. He is his own independent police force.

O'Shay does not pull a shift, he simply operates until he drops. He does not have a master plan, he has a hunch, hears a rumor, feels a slight tickle on this web he has created and these impulses that course toward him are assessed and measured. There are no big cases, not at first, there are simply signals from the streets and some of these things lead to other things. A phone call to a snitch and then eventually Colombia walks through the door. But there is never really downtime, what looks like waiting is still the work.

In the big office where he tends his voodoo altar, agents have meetings and projects and big schemes. They are clerks to their own hallucinations. O'Shay is days and nights of little things, small snatches of information and gossip. He is the predator sniffing the wind and stumbling on the game trail.

O'Shay needs that sensation, everyone does, though few ever taste it. For years, his DEA supervisor would come by the office late at night, look in and see O'Shay toiling away all alone, head down, files open, phone cradled to his ear, and think, Ah, this guy will make my career.

From time to time, wire intercepts would pick up threats against O'Shay's life and according to the protocol of the agency, this meant O'Shay must be removed from his cases while the threat was investigated and also that he must be told of the threat to his life. This was never done. In part because removing him from his work would derail his cases and his cases were helping to make his supervisor's career. But there was a second factor, the absolute belief that if O'Shay were told of the threat, he would go out and kill the person who threatened him. So a former head supervisor admits he told him nothing and let him continue with his work.

He is electric with his work. Others wait, others are bored and they experience such moments because of their own emptiness.

O'Shay is certain of this fact.

O'SHAY IS SLIPPING INTO MEMORY. And this unnerves Bobbie. The past must be a separate country, she knows this, it is the key to sanity. But of late, he has gone into the past. She only glimpses this, and she has to wonder how he is with others. Or what he is now thinking when he is alone.

O'Shay is staring straight ahead as the car rumbles down the black streets, his face a faint smear on the glass from the dim lights of the dash, the bulky body a silhouette. A light rain walks across the city, the tires swish through the sheen of water, the windshield wipers sound like a brush caressing the cymbal.

"What's your favorite book?" he asks suddenly.

"*To Kill a Mockingbird,*" he answers for himself.

Out of the blue he muses that he's only dated a few white women in his life.

"Raised with other races," he says.

He's trying to lose weight. He slowly sips iced tea. He sits alone on a restaurant patio. He does not eat. A mockingbird lands on an empty chair. He examines his gun, all the surface bluing worn off because he wears it tucked inside his pants, which is dangerous since the gun can accidentally fire. The mockingbird watches from the empty chair as his hand caresses the weapon. The greasy aroma of hamburgers on the grill drifts out from the restaurant and envelops the patio. O'Shay is lost in the wear on his gun, the mockingbird sitting without fear two feet away, the empty virtue of the iced tea sloshing in his mouth.

He is lost in thought and yet totally conscious of his surroundings. He can sit erect, be a block that will never move. He can slouch down, lean forward, his large forearms on the table, a faint smile on his face, and then shrink, the body slack and at ease. He can do this in a second. He is always on.

The shootings happen in real time, he thinks, none of that reported slow-motion stuff. Maybe it seems like slow motion to some people when they relive it later. But the gunfire itself, he never seems to hear the loud bang of the round. As they pass his head they go whumpf, whumpf.

That first shootout where someone died, he was twenty-five.

"Supposed to fuck you up," he knows. "But I didn't feel anything." He admits that he's supposed to feel something. Then he falls silent a moment and adds in a flat voice, "I'm a predator who hunts other predators."

Bobbie is also sliding into the past, resisting the pull mightily, but nevertheless gliding into the mists she thought she'd left far behind.

"Here," she'll say, "take a look." And then she rolls up her sleeve on her right arm to show a canyon of scar tissue gouged deep into her flesh. Her skin has never seen a single ray of sun.

"I like to shoot the coke," she purrs.

"It's not the same if you snort it," she explains. "I love to shoot. Look, I hold down two jobs, I pay my bills and taxes, I've never been on welfare. It's a lie that the stuff makes you dysfunctional. Why, sometimes I shoot up driving on the freeway."

And then she pauses, takes another hungry drag, expels the smoke in the same elegant manner, creating a lacework in the still, wet air.

"Don't tell him," she adds. "He doesn't know I'm still shooting."

She looks out at the bum in the parking lot and says, "I hate them. I pay my bills."

Her windowless office nestles downstairs at the hotel. The office is small, the fixture a color-draining fluorescent which makes her pale skin even paler. She keeps a list of all the Mexican and Honduran whores who use the hotel. The names are normal, Maria, Guadalupe, Josefina, and so forth and the ages run seventeen, eighteen, and nineteen. She is not supposed to have such a list but she has a need for things she is not supposed to have. Just as she has a need for quickness—her fingers will fly across the computer keyboard and instantly accounts appear and disappear as Joey O'Shay needs space for guests who exist and yet must not really exist.

The years seem to have steamrolled over her good-time days with the Dixie mafia. But her eyes are quick and sharp, she has smart eyes. Her tongue is swift, also. The city bakes with a sense of waiting for something to happen that will rend the vast limpness in the air.

―――――

MAYBE IT'S IN THE BLOOD, he thinks. Joey O'Shay moves in on thirty, he is growing a world in the streets, and yet this feeling gnaws at him, this sense of something not being fed by his life. He wants beauty, color, form, something that whispers magic to him. He has honed his abilities with a gun, with an ice pick, he can read the souls of others by a glance into their eyes. He knows his market and its needs, he's plowed through books on chemistry and the brain night after night to comprehend both his products and his customers. And most important, he is rising from where he came and now stands on ground he never even dared to dream he might reach. He's in a crew, his life is action, his body is hard and swings through the people around him like an ax.

Joey O'Shay and his crew go into the photography studio, the kind with props and costumes to propel the clients into another time and place. There are five of them, all bearded or with mustaches, all wearing poker faces lit by the glare of their eyes. One wears an Indian bone necklace, half a dozen strands thick as a choker, a Winchester 94 in his hand. Another has a boiled shirt, a Stetson, a marshal's long coat, and that star pinned on his chest. The black guy is a buffalo soldier in a blue coat, boots with his britches tucked in the top, gun resting on his lap. The last guy has on a derby, corncob pipe in mouth, double-barreled shotgun, black vest, and a studied gaze. No one smiles, not a flicker of warmth.

O'Shay is off to the left, the only man not standing or seated. He's on one knee, rifle held in right hand, muzzle down, the big knife in his left held low, held as if caught in the instant before it rips upward to the gut. He is the only man not in costume, save for saddlebags leaning against him.

The photo pleases them, a portrait of cutthroats. They are young and foolish and willing and now they have a record of their lives.

———

HE DOES NOT CARE MUCH for clothes, and cars are simply machines to him, sometimes useful to advertise what he is, but still, nothing more than devices to enable him to move. But he hungers for beauty. The green and dark and light along the creek, the sun bouncing off the still waters of the lake in the morning, the flash of a bird's wing, the scent of food in the air made solid and brilliant and actual on a canvas.

He remembers beauty, inches his way back to that point in his life when the grace of a girl, the smear of the sky could stop him in his tracks. He reaches back and remembers his grandfather telling him of his own mother, a woman with some kind of culture, a refined person who could play a piano. And paint. She died on his grandfather when he was a child and he carried that death like a wound on his face till the end of his days. The old man would tell him of her paintings, talk of them as if they were the Rose of Sharon, beauty beyond comprehension or even belief. How they'd hung in the house shining down on a grand piano and then the house burned and they went with the fire.

Maybe, just maybe, some of her is lurking in his blood, that's what O'Shay thinks as he rolls through the city doing his deals. He starts hanging out at art shows, and soon, given his business it's not hard, he knows some artists. They are always intrigued by his kind. He devours piles of art books in his spare moments. O'Shay is incapable of doing anything in a half-assed manner.

He starts playing with various mediums and techniques, feeling his way into a sense of color, ambling steadily back to the creek when everything was electric with color and scent and form. Yet oddly, in his memory, these sensations are almost abstract, as if life had escaped the prison of being a tree or a bird or a turtle or a breeze.

He forms a loose habit of staying up very late at night, staying up after the work of the streets is done, after the haggling over kilos and prices has passed for the moment, staying up alone with a canvas, soft music playing or some jazz, standing there with a brush, the colors, and the invitation of the blank form. And he is invited into his feelings, into his dreams and memories, he goes happily back to the beauty thing. Maybe, just maybe, he can get down something that lasts in a life where everyone around him perishes.

At these moments, he feels alive and alert and yet relaxed because in the full groan of the night, he seems to escape himself and become the colors blooming before his eyes. He is there late at night with his great-grandmother, a ghost he knows only from photographs, the fine-looking lady who made beauty while living in this hard world. Jazz purring, the house silent but for the movement of the brush, the beauty thing, the feelings. He guards these moments. They are dangerous, they are an indulgence.

But he cannot help himself. He needs this beauty thing and so he feeds it despite the danger.

SONRISA AND O'SHAY approach a fucking punk, maybe twenty-something, just a little shit. The night has settled over the city. They are at ease, or at least O'Shay is. He's become the muscle, the pistol that follows Sonrisa into the deals, little deals that emanate from houses. This is a more innocent time in the drug world, before the major weight flooded the nation. The roles are not written on a stone tablet, they are simply the way it plays out. And so O'Shay cues himself off Sonrisa's acts and manners. He lets him create the play. And the deals keep coming.

It is dark now, they are going to a deal, and this punk is in front of them, a little nothing and not a concern. O'Shay is alert but still relaxed. He is mass, the kid is nothing, a flyspeck in their way as they go to do a deal.

Suddenly, Sonrisa's gun comes out of nowhere and he draws down on the kid.

O'Shay shakes him down and finds a .22 Magnum.

It is that fast.

That is the point for Sonrisa.

That is the lesson: anyone can be death for you, even a punk standing on the street, just part of the landscape that lies outside your deal.

When you sense danger, instantly move, fuck reasons, don't sort it out, move, and if you must, kill, but for God's sake don't pause, don't ponder, because you get only one serious mistake and then it is over.

The reason for the move, the reason you act, the signal you have picked up, well, you can sort that out later. Maybe you will never understand what you sensed. That does not matter. But paying attention to this feeling, this sudden warning shout within your body, that does matter. And if this shout is loud and raw, and you bring that gun up and fire, then shoot to kill, empty the whole fucking clip except one fucking round, don't try any finesse, forget marksmanship. Obliterate the thing that threatens you.

He has learned that the best backup is your willingness to stab someone over and over and over, to be hyperaggressive.

HERE ARE SOME RULES, like all rules made to be broken. Never use product, that will take away the edge. Never relax, that will let danger near. Never get close to anyone, that will destroy judgment. Never share information, that will cause death and ruin. Never trust, never, never trust.

Never drink to excess, or when the moment comes that drink takes over the senses, be in a safe place. Never believe there is a safe place because that place will become a tomb.

Never hurry, no deal is ever better than the deal not done.

Never make a mistake, eyes are watching.

And enjoy the wait, savor it, cherish it, languish in it.

The wait is the test. Impatience is a tell, a clue to others, blood in the water, a signal for a feeding frenzy and destruction.

Tires whir down the night street, the wait continues, and what is your favorite book?

O'SHAY PULLS OVER to a curb, beckons, and the girl comes to the window on the driver's side. She says she is clean, she's been clean three years, but she has this problem, that fucking bitch that monitors her, well that cunt is going to violate her over a urine test, and she's fuckin' clean.

O'Shay looks at her, the eyes, the skin, and says nothing. She brings him things, little bits of this and that, words floating down streets. He remembers when she was younger, fresher, before her brother fucked her and she went to junk. He has a roll, peels off a hundred. He promises nothing but she knows he can reach into places he will not admit.

She tries again, trying to get some kind of guarantee. Yes, she remembers, she was out on a date with a guy, a nice dinner, he's moving big, she says, and over dinner he tells her he took a wetback out south of the city, beat the shit out of him, then burned him alive.

O'Shay stays silent.

She backs off, knows her time is up. Then her father approaches, the face sallow, the skin past fatigue. He clutches a bottle of water to fight back the feel of his diabetes. His hair is falling out, he can barely stand. His life has often spun on a needle. He whispers to O'Shay, head leaning forward so they almost touch, whispers quickly. Another hundred.

And then he fades back.

O'Shay pulls away, trolling for nothing, just keeping in touch while he waits.

———

WEEKS GO BY, the Colombian connections stall, there are excuses, problems with phone calls, the need to get the right stuff together, arrangements to be made. The wait.

Few words. In the business, conversations are very short and coded and clipped—a friend...suitcases...cartons...delivery. They come like darts whirring through the air, hit the board, and then action ceases once again and the wait continues.

The city rolls on, the towers shine, days come and go, the wait flows and becomes the deal.

COSIMA STARTS ANOTHER PAINTING. This time the sky is orange and white and yellow and the sun looks like it is red, white, and blue. The eagle flies, beak open, wings out, talons extended.

In the talons is the severed head, that filthy beard, the turban, talons digging into the skull, a trickle of blood spreading on the turban.

This is better, she can feel it is better.

THE WHITE TABLE SLUMBERS in the damp air of the patio. They have a mixed platter of oysters, crab, shrimp, clams, and scallops, all deep-fried for dunking in various salsas.

O'Shay sits down next to Jaime, who wears a black shirt, black trousers, black boots, and thick gold chains. Jaime moves back and forth, crosses the line and parties with the big dealers, comes back and runs his business. He is into trucks. Years ago he went down and lost a million or so but now he's in fine shape, has done his stretch and is back in business.

He's explaining Viagra, how it works fine, but hell, his head felt like it was exploding from all the blood, so he allows it is a mixed bag.

Jaime chatters, he laughs, he is the animation at the table, a flurry of fast talk and shifting subjects. He is always working and part of his work is to keep everyone else off balance. He eyes O'Shay, can taste something going on in him as he sits passive at the table barely eating, seldom speaking. His shoulders are hunched, his head bent down, his big forearms rest on the table, the hands clasped. O'Shay wears not a speck of jewelry. He is simply work and calm and Jaime can tell at this moment in the midst of O'Shay's calm is tension and he can smell in this tension a deal.

Jaime wants into the action. That is part of the business, always reaching out for new bits of deals, always having more than one thing going, never counting on anything, never pulling down the shades and locking the door and saying the business is closed until regular hours tomorrow. In the work there is no rest, only pauses, and if the pauses continue for too long, then the business is over. So Jaime reaches out, seeking a taste of the action.

He has an idea to sell. A friend works in the rich-man's club atop a tower downtown and a lawyer came in last week and asked his friend if he knew anyone who wanted a thousand kilos of coke.

O'Shay instantly brightens.

"A fucking lawyer?" he says. "I want him."

Suddenly Jaime is on the phone, talking softly, sending out feelers for a meet with the lawyer. O'Shay sits quietly, amused by the notion of a rich boy in one of the towers trying to make a killing in the business. Amateurs are always a diversion as they mimic movies and act out roles written for ninety-minute lives on a screen. He can see the asshole attorney boning up on flicks. And he knows the greed will be almost palpable. Why else would a straight try to do such a deal? And greed, like any desire, is weakness.

A harsh remark about the FBI drifts across the table.

Jaime is enraged. He is an immigrant. He loves his new country, and the FBI especially.

"I would die for the FBI," he says.

O'Shay says nothing. He is always the same. The face alert, yet blank.

BOBBIE LOVES HER ROUNDS. She prides herself on doing everything better and faster than anyone else can. She hates waiting for others as they muddle through.

She's on a tear about unions.

"That guy I just dropped off," she says, "how much does he make? Well, I fucking know. He gets a hundred and ten large a year and all he knows how to do is screw sheet metal together. Unions, fucking unions. I hate them. They don't earn their keep, they don't compete, they're parasites. That's what is wrong with this country. Unions."

O'SHAY IS PARKED by Bobbie's hotel listening to music and lost in thought. The voice seeps through the speakers, "...tell me about the dreams...tell me about the deaths...anything you want to know, I'll tell you."

It is two a.m. dead, the lap-dance joints up the road are ending their pleasures and nothing seems to be moving but the homeless searching for parked cars to rob. O'Shay slumps forward, his arms cradled on top of the wheel. He wonders what dreams feel like, he wonders why anyone would want to know about the deaths, which are always more private than any moment of love.

He can hardly remember the women but he always remembers the killings.

He starts his car, he pulls out of the hotel lot. His night is not over, but he feels the need for movement. The wait continues.

The lawyer with the thousand kilos, he's in play. He feels safe, everyone in clubs atop big towers feels safe, immune. O'Shay is safer. He knows the wait. He knows patience. He does not know immunity, only the losers and the dead know that feeling.

IT IS LATE IN THE AFTERNOON, and O'Shay is hell-bent on finding a church whose style he admires, a simple wooden church, the kind thrown up a century ago by the poor, as a gesture toward their God. This interest, like his paintings, is something he keeps secret from others and often from himself. He is sensitive to color and form and structure and lines and smells and sounds. Music bathes him, "Come on down where angels cry, they don't cry for me."

He snaps off his sound system. He stares out his windshield, fuzzy as to the exact location of the church he seeks. He will not go in, he is sure of that. Church never seems to help him. Mass is long and dreary, the sermons false and loud. The choirs, well, they are hopeful but they are fools.

He inhales deeply, tastes the green off the trees, the fumes in the air, the hint of barbecue off some joint a block away. He hears birdsong.

O'Shay goes up and down side streets in the hard neighborhood by the freeway. The trees arch over the road and make a world of green. Young girls stand on porches with short shorts, skimpy halters, a can of pop, and blank faces. O'Shay finds the church finally. The lot is empty and freshly bladed. The city eats a bite of its past. He is disappointed, as if someone had picked his pocket.

A whore stands on the corner in the sun. Her body is rail thin, her hair lank and dirty, and she has no hips, just bones poking against her capris.

He pulls over, asks her about the church. She stares and says nothing.

THE PLANET IS PURPLE, a lovely purple and it is surrounded by bands, a yellow band, a green band, then a yellow-green band, then a green band again, then a faint red band that looks like blood drying. Cosima leans back, takes in her work, puts down the brush, and thinks.

She is tired of waiting, but more than that she is angry at being left outside the deal. She knows O'Shay is doing something big, she can sense this even if he thinks she is in the dark.

He needs her and he will learn this when she finds her way into his deal.

She picks up the brush and paints a fuse on the top of the planet. Here, yes, a dab of red. Now it feels perfect.

The fuse is lit.

THE COLOMBIAN DEAL will work because Joey O'Shay does not fail. That is the problem for him: the emptiness of winning. If you plunge deeply enough into the business, you lose all sense of boundaries. You leave the place of simple right and wrong. And when you leave that place, you are not in a strange place but in the place where you have always been. But now it looks quite different to you.

O'Shay sits and waits so that he can enjoy the wait and want the wait. He tells himself, "I love swamps, creeks, oceans, I love water. I love floating."

So he enters the float. He puts a stack of CDs on, soft music. He thinks this stuff is shit. But it helps the wait.

Candles burn on his altar.

The black coffin sits hungry.

He thinks of a song his son wrote about him. Yes, suddenly that song comes to him. He keeps a copy of it in his truck and sometimes, when he is alone he plays it. He hates the song and yet he loves the song.

Now the song plays in his head:

> She pours a glass and drinks about you
> And dances with memories inside her head
> She'll dance alone, slow without you
> She'll hear the words you never said
> Don't you know that

BOBBIE TURNS UP LATE in the day. She is in good spirits, it is her destiny.

She thinks about the hepatitis C. They've given her a year or two. She doesn't complain, what the hell, all those needles catch up with you, she explains.

She's been to doctors. And counselors. They all want her to stop drinking and stop shooting coke.

She snorts with contempt.

"This makes me happy," she notes. "Why would I give it up? I mean what's the point, you don't drink, don't use coke, then what?"

The legal drugs that keep bad things at bay, well, she must clean up to take those and then she must hand over a lot of money. Liver transplants, there is a list and rules for that.

She's already checked.

It's simple: if she keeps drinking and shooting she will most likely be dead in a year or so.

The night of that dream, the one where she is trapped and O'Shay must come and rescue her, yes, that night, the dream ended suddenly in a stab of pain. She sat up in bed and felt her shoulder and that was the first moment she noticed the tumors.

She must be inside O'Shay's new deal. That is the only way she can stay alive, she thinks, or even feel alive.

———

THE CAPITOL GLEAMS WHITE in the night, the dome, the columns holding up the dome, and between the columns a blaze of yellow light. The building is alive in the painting, and looks warm and beckoning under the dark blue-black of night. Now she paints the envelopes, white envelopes, yellow envelopes, and out of the bottom corner of one falls a white powder, a soft cascade down on the Capitol.

The statue atop the dome is severed below the waist by the top edge of the canvas. This does not matter.

It is the Capitol that matters, the safe glow of the light within, the powder falling from the envelopes and raining down death.

Cosima hates them.

BOBBIE HATES STORIES THAT WORK because she knows they are lies. Real stories, she thinks, refuse to follow the script, they will not bend and follow the plot. They deny redemption. They do not address problems. The Bible will not help in these stories, neither the Old Testament nor the New. The book of Revelations reeks of bad acid, the Psalms are limp-wristed. And Exodus goes nowhere, a promised land turns out to be not much at all. The cross, *that* you bear, but the stone does not roll away from the mouth of the tomb on the third day. This does not mean you give up hope. Quite the opposite. It means you are really here.

"Look," she says, "I can get a liver. But then I can never drink or do coke. Not ever. And besides, I gotta clean up to even qualify for the liver. But forget that, suppose I do clean up and they give me a new liver. What then?"

She almost beams. The bar is a nice one in a classy hotel, the kind with old wood and sofas, a place where locals are not welcome. It is happy hour, and for a spell, people have been drifting in and greeting each other. A business meeting.

The guys all go up to their rooms when they check in, dump

their gear, and then come back down in the same tired slacks and wrinkled shirts. Many of them have faint pale lines on their fingers where rings normally rest.

The women are quite different. They come into the hotel, glance over at the bar, and come right in, wheeling their luggage behind them. They shout greetings, make quick and careful embraces. They all wear neat, dark suits, and lipstick but little other makeup. They sense the possibility of a fling, manageable, but still a break in routine. They will not drink out of a slipper after dinner, but they will put on that black bra when they finish unpacking and come back down to the bar.

Bobbie's good at this and she knows it. She has always seen it coming, well, almost always. She has made people at a glance, pegged cops before they opened their mouths. Sensed a room instantly, managed her affairs, paid her bills, taken in the news. To her the hotel is a parade of the obvious—God, she thinks, people fucking broadcast who they are and what they want and what their weaknesses are, they broadcast their weaknesses simply by announcing who they are and what they want.

Bobbie watches the parade now but it does not interest her. She insists on living on another level. The straight world has retirement plans, regular hours, a recommended trajectory. This trajectory dulls the appetite, proceeds by slow clocks, and leads to the same place that the rush in her veins leads, to death. That's it, it is not about a destination, it is about the ride.

She is still beaming as she thinks about her question.

"Here's my question," Bobbie purrs. "What good is a liver you can never use?"

HE'S WORKING VICE in his early days, the whores, the bathhouses, which he thinks of as "all that sick shit," he's in it all because it is a living. The women, Jesus, he wonders how anyone can stick it in them since they all have purses that rattle with pills to keep at bay all the bugs and whatnot raging within them.

The homosexuals seem to slink around but there is real money in the bathhouses. He faces the variety of human appetites. The ministers catch his eye, especially the damn Baptists with their huge pompadours. The whores, some of them, he kind of gets close to, understands the hungers that drive them to such places. But then he understands more than he likes to admit. He also understands appetite, feels it in himself and disciplines his desires, keeps drink at bay, avoids all drugs. Hits the iron, runs the streets at night.

But what he learns working the skin trade are the early secrets, the way to feel his way into someone else's hunger, to catch that glimmer of someone's weakness, the way to play people and bend them or break them to your will. He starts recruiting hookers, some of them teenage girls. He builds and reaches out and bit by bit he enlarges his world and his eyes and his ears. He smells everything that moves.

Walk down a hall, the door is slightly open, a man is on all fours, dog collar around his neck, the woman in high heels holding a black riding crop. An hour later the man comes out of the room and now he looks just like a lawyer as he returns to his work and public life. Appetites that can only be fed by outlaws and that for this reason cost real money, markets to be followed and exploited, secrets to be learned and used. A crash course in psychology in which he enrolls to learn not how to heal but how to control. He learns there is a place to go, a place outside of self and deep into someone else, a place where all the senses open and torrents of feelings pour in and you are that someone else, his needs and actions are all predictable. Joey O'Shay learns how to savor those moments and through them he takes over other human beings.

He notices that people watch from the outside and look for a move or mistake and then pounce. But he learns that if he watches from the inside, from inside the person, he no longer has to make calculations, he no longer sits waiting for some key moment, he is always in the moment and in control and he does not have to pounce, he becomes the author of the other person's moves, even moves the person has not made. Because O'Shay knows what they will do in places where they have never been.

He lifts, runs, glowers but he controls with something else and he cannot find a word for this something else except a kind of feeling. Facts become slippery things, not like they seem to be for others, not touchstones or absolutes, but simply signals to a deeper level, a level where things hop and skip and connect in ways that logic pretends is not possible. He'll pull an eighteen-year-old hooker into his orbit and the facts say she is a whore and a junkie but this other sense teaches him that she is a hunger, a person, and that if he reaches that hunger and person she becomes trustworthy and dutiful and useful and the categories and ways other people use for filing experiences make them blind to this possibility. But his eyes are open to it.

So in the steam of the bathhouses with the naked men moving in the half-light, and in the eyes of whores leaning into his car window when he pulls over to talk with them and check out the night, he becomes keen on this other level and gradually he moves there. He tells no one of this move, he simply makes it. And once he has moved, he begins to make his drug connections through the people he controls and breaks and bends and does whatever he wishes with. That is how it begins for him. He finds people who will do anything to get high and finds he gets high by finding them and using them.

He is married now, he has fathered a child, a boy. He has the steady job with benefits and security, the stuff of American life. But he is his work. In his mind, there is no office he goes to in the morning or leaves in the evening, there is only a calling. And this thing, this kind of work does not simply devour his normal life. It becomes life itself.

HE'S A BEAT COP. He enters the brick house around four a.m. He passes a woman in the hall and she says, "He's back there."

O'Shay moves through the darkened house. All at once, a man comes out of a bedroom with a .357 and strides into the hallway. He is chest to chest with O'Shay. The man's gun comes up, fires, and the bullet passes close by his head. O'Shay instantly pumps two rounds into the man's chest. He falls dead to the floor.

He is twenty-five and he has had to kill a man. As he travels down the hallway and out the door, he passes the wife.

"I'm sorry," he says.

She nods.

When the man lurched out of the bedroom with the gun, O'Shay thought, "I'm not going to die in this shithole. I've got a young son," and then he could smell the scent of his child's hair.

There was blood everywhere. His ears rang. For a while, a thought would flash across his mind: I was almost killed.

This passed.

Afterward, he finds he does not like to talk about the killing. He hates thinking of "blood falling out, really sick shit."

And then a week later, he is in another shooting. It does not seem to bother him. He does not know why, but still, it does not seem to bother him. He lives in some state he cannot describe. Sometimes he'll see movies and they'll have everything he is living in slow motion and that is not the way it is for him, there is no slow motion, his life does not pass before his eyes, the bullets, he feels them pass by his head, a kind of whooshing sound, he acts and shoots but no slow motion, time is very, very real at such moments.

He joins a crew made up of the kind of guys he knew in his old neighborhood, hard cases basically, some of them vets, all of them looking for a place in the city, a position of sorts that will feed them. O'Shay is usually point man in their adventures, along with Tye and Red, his two black partners (who never left his circle even after they transferred out), but he is the first through the door. They start doing deals, maybe one a day, and if they must, they use guns to do their deals—.357s and shotguns. Because of the appetite for rip-offs in deals, it is a rough and quick school.

He is surprised by the feel of things. Logically, he thinks the first guy through a door, generally himself, is the most likely to take a bullet. But it does not happen that way. People around him and behind him go down and he remains unscathed and he wonders about this and then does not wonder because thinking may not be healthy in such work.

He enters a room, the only light is the glare coming off the television, he blows out the screen with a shotgun, the firefight takes him to his pistol, rounds roar out in the room. He is standing at the end. He does not know why. He cannot even yet ask the question.

But no talk. He dreads certain things: of talking shit and seeing that look in the eyes of others, of talking shit about the killings and disrespecting the dead. There is a deeper zone, the place, perhaps, where a blues guitarist finds choice licks, the place, just maybe, where a poet finds the words, the place an athlete reaches into for the moves. And no one who has been there wants to talk about this place. They say it's hard to find the words. They say they don't really understand it themselves. They say almost anything but the truth and the truth is, they don't think other people deserve to know unless they've paid the price to get there.

The real question, he thinks, is why some die and some live. Drive down the street and the faces say what you need to know and the messages can be taken down in a glance. The feel of the gun, the body as a tool, the penetration of other lives.

A drink, two drinks, maybe sleep will come.

...you get better and then you start feeling like a ghost. You are not really there. You put ego aside, and you do it quickly and quietly...

THE HAIR IS A BROWN MESS of tassels hanging down and defiant of the brush. The short beard frames a face empty of expression except one: What do you have? The eyes look out but slam shut at the same time and are as empty and withholding as two garage doors. The face gives up nothing but asks and probes and invades others. The hands are large and firm in the photo taken with the crew, one holds a knife, a hunting knife with a good blade for gutting. The other hand wraps around a long gun.

The skin on the face is smooth, the skin of a child, skin without guile or worry, the skin that is supposed to come from some inner serenity. The skin is the ultimate camouflage of Joey O'Shay. When women see this face, they wince and they open up, they see hurt in the face, they see caring of some sort in this face, they sense vulnerability. Women fall in love with this face.

Men do not. They see the dead eyes, the relaxed skin, the lack of tension. Men see death in this face and they move on and do not want to discover what is behind this face because they already know and fear what they see.

When the face speaks, and it does this no more than is absolutely necessary, the words are flat, almost soft, and measured, the words never stumble or rush and there are no extra words. The words never pause, they simply march out of his mouth. They do not coax. They state. The face says always the same thing: Take it or leave it. But do not try to change it.

All this from a face that is perfectly blank. Ageless, it seems.

———

STRANDS OF SOFT COLORS streak the sky, blues, yellows, pinks, reds, and they all shout danger and warn of storms. In the foreground, rows of missile-shaped objects faintly loom, women with veils, draped in light-colored garments. Cosima, drenched in scent, her lips bright with lipstick, looks down at her work.

She has been painting her terrors for months now, almost in a rut of hatred and superiority. She has found the ultimate target for her contempt, something so anti-life it makes her feel alive, and almost whole.

At the back of the rows of shrouded women stands the bearded man, and he has that smile, not of saintliness but of being in power and the boss. He smiles like all the men Cosima has destroyed in her life.

And behind him is a brown and yet spectral form: a fedora, no face visible, a brown suit and waving aloft an AK-47, a tool of the business now profaned by its use by such enemies. It is for Cosima a blasphemy, this use of the AK-47 for such black work.

But what really attracts her to her painting is the unfolding of power: the helpless dronelike women veiled in the foreground, the smug, smiling man who thinks he is the boss, and behind them like a brown mountain, the real man pulling all the strings.

The painting suggests the feel of Cosima's own world, with Joey O'Shay always just out of plain view and yet always picking and determining the action.

THE MOVIE WILL NOT STOP RUNNING. O'Shay cannot reach over and hit a button and make it stop. She has thick lips, a childish face, one not hardened by life. The guy has been choking her now for almost fifteen minutes. She persists, in those brief interludes when he removes the clear plastic bag and lets her breathe, she persists in thinking this is punishment, or for the man some kind of thrill but this is not real. She is not being slowly tortured and murdered.

The room is drab: a sheet on the floor, her tiny sixteen-year-old body pinned by the older and larger man, a blank painted wall, a carpet that rolls out from under the sheet, that is it.

No furniture.

No pictures on the wall.

No memories. It is all very modern and stark and clean. It is pure.

She has dark freckles here and there on her face and arms and legs. Her clothes are rumpled but that is understandable, just as her hair is increasingly disheveled. Her hands flail, just as her hips and legs writhe, but she does not grab his hands or strike at his body. There is an acceptance of pain.

There are phrases—the heart of darkness, the depths of evil— but she does not know this language. That is clear. She is used to being hurt, being struck, being fucked, being beaten. She accepts.

On the street, she hooks, a sixteen-year-old whore known as Popsicle. The man choking her is older, in his late twenties. He sells drugs, he has an apartment, and she is in that apartment now.

So she accepts even as she begs.

She knows pain is part of her life. She pretends this is simply more of the same.

But as the minutes go by, as the sessions of strangulation grow longer, as the man and the person running the camera fail to respond to her pleas, she seems to begin to wonder. Her kicks and writhing grow more animated.

He has been at it now for fifteen minutes, perhaps even a little longer. He gives no sign of being satiated. His appetite seems to be growing, not diminishing.

And there is something subtle but real in the camera work: it feels feminine. The movements in and out do not so much capture action the way a male would, that obsession with the money shot, no, the camera seems to go into almost a trance when she is choking, the little legs thrashing, to become still as oxygen drains from her body, and then suddenly, when she seems to be all but dead, the camera will move in to capture her face through the clear plastic.

The camera has the feel of a woman who is not into action but tastes the emotion of the moment. The camera shows a kind of caring, not love but engagement.

HE SITS THROUGH THE NIGHT trying to remember who he once was. In his teens, he ran with the wild boys.

His friend's mom runs a bar that is the clubhouse for bikers and local mafia souls. She had eight husbands, six kids by as many men, and one of her husbands burned the hell out of the friend's legs, out of simple cussedness.

He and O'Shay were made for each other. They broke into warehouses and stole things.

One night when they are sixteen, his friend asks O'Shay to give him a ride to a downtown bar frequented by homicide detectives, narcs, and criminals.

He says simply, "I gotta find my daddy, he beat my momma bad."

The night is cold, and the bar is clouds of smoke, a herd of lowlifes and their bitches, and on the long wooden bar a midget woman is dancing to a rockabilly song off the jukebox.

"What the fuck are you doing here?" a guy with greased-back hair asks O'Shay.

"I don't know."

That's how he grew up—fights, theft, bars, dancing midgets, putting in Sheetrock at night for two fucking dollars an hour, buying stolen clothes that were ugly as shit, goddamn white-trashy clothes and somehow in that stinking bar Ray Charles is always playing.

Ah, his old neighborhood, he can damn near smell the place now, faces that are black, Italian, Greek, Mexican, poor whites and walking home at night food in the air, onions, grilled meat, cilantro, white boys learning to cook Mexican food, black boys learning to cook cabbage and corned beef, a fucking human stew of a place that produced, well, O'Shay thinks, that produced people just like himself.

Drunk at night, raising hell, Christ, he was thirteen, maybe, first time a cop worked him over, whomped his head against a wall and braced him good. Or cops would catch him drinking, take his booze and whip his ass. Sometimes haul him to a gas station in the winter, make him lie on the ground while they hosed him down, and he'd be freezing there on the ground and they'd say, "Hey, you can't get up until you count to a hundred out loud." It became a game both the cops and kids loved.

His parents are gone, working out of state. He and his brother kind of keep house and start running poker games at night, and O'Shay cooks food, skims some money from the table and there isn't much but there is always a freezer full of meat because his mother came up hungry and that kind of hunger you never forget or put behind you and one night during a poker game a Chicano gets hungry and at the bottom of the freezer finds a frozen redbone hound, his mom's dog that had gotten run over and that she'd put on ice until she got around to a proper burying and O'Shay thinks, that's the kind of off-the-wall family he comes from.

She is olive skin, dark eyes, and long lashes that look like butterflies. He finds where she lives, a big country house with a porch wrapping around it and smothered in honeysuckle, the sweet scent floating out and a huge ancient oak in front. His first love. When he talks to her, his heart pounds so hard he thinks it'll explode out of his chest. Sitting on the porch with her on the swing, talking about this and that, the day was twenty-four hours then, not like now, but real time, quality time, the light falling down on the green leaves.

Years later, he runs into her at a school reunion. He's got long hair and a beard, and she's exactly as she was, never married, butterfly eyelashes, and she tells him she still remembers the lime scent coming off his letter sweater. And suddenly he remembers her scent, a clean thing like some magic lotion and everything is clean when he remembers that scent, the entire world is clean and possible and he is back with the honeysuckle on a swing with his girl.

He is at the bar, he is a kid, has downed three or four beers. It's Friday and two white guys in the corner of the bar seem to be having an argument and O'Shay hears something go pop pop pop and looks over and sees one white guy on the floor lying on his gut.

This woman is leaning over him asking, "What's wrong?"

And the guy says, "Why, you stupid bitch, I'm shot."

Then he dies.

The cops come, everybody lies, O'Shay and his friend run.

FOR SIX WEEKS, O'Shay pushes the connection out of his thoughts. Cosima talks to Alvarez occasionally but the conversations are short. And coded. Talk about boxes, shipments. Business.

Alvarez comes back and they go to the apartment, a dull cell in a huge complex. O'Shay likes nondescript venues for his business. Alvarez has brought a sample half kilo, a testament to what commerce they could do in the future. O'Shay listens, one gunman hanging back in the room, Cosima fluttering about like a maid. No one talks but Alvarez and O'Shay.

The pornography business matters to Alvarez. He sees online porn services as an ideal way to launder money, a kind of ultimate restaurant where receipts can be dummied forever to explain income and where the details of cooks and food and storefronts can be avoided. O'Shay explains various options. Alvarez is keen for a webcam where someone can call in and direct the woman to do certain things on camera. He asks if O'Shay would like to invest.

"I don't need heat like that," O'Shay tells him. "It may be all perfectly legal but I have my businesses already. I have the trucks and the planes and I don't need the fucking feds coming around because I've got some bitches fucking on camera."

The talk goes back to heroin, to shipments, prices, purity. Transport. O'Shay is meticulous on such points. He hates sloppy work, he worships punctuality. He insists on sound habits from anyone he deals with. So he drones on about the details, about how the heroin will be packaged.

O'Shay says, "I know how you bring it in, stuffed in someone's gut and blasted out their fucking ass. But I do not ever want to touch this stuff until you've cleaned it up. Understand?"

"No, no, no," Alvarez interrupts. "You will never ever have that problem with me. It will be clean, my workers will clean it up. And you will not pay for the wrappings, they will be peeled. It will only be the sticks of heroin and they will be immaculately clean."

Suddenly O'Shay shifts back to the pornography business, starts talking about other possibilities there. Alvarez listens attentively. That is the moment O'Shay's aide leaves the room.

When he returns, the aide shoves a machine gun in Alvarez's face.

He is ordered down on the floor like a dog and he obeys.

He is forced to lie flat and spread his legs, put his hands behind his head. The cold barrel caresses his skull.

Alvarez is experienced in handling many things at the same time. As he sprawls on the floor and smells ruin and death nearby, he must consider other matters. He is involved in a scheme to murder a federal judge, and this cannot be escaped. The judge's fate is merely a chip in a larger game. A large Colombian player is in prison and wishes a way out. His people will arrange for a rival organization to murder the judge, he is told. Alvarez is to be with them, to help them set up the hit, to find them a place to stay. And then vanish and call his Colombian colleagues. They will sell the killers to the FBI, help prevent the murder of a federal judge. All this for a sentence consideration for their boss in prison. It will be a clean thing, and a thing Alvarez with his lost loads cannot possibly refuse. He smells the carpet that scrapes his face, feels the gun at his head, and also thinks he may fail in his assignment concerning the judge. He remains calm.

O'Shay does not rise from his chair but he watches attentively. He notes that Alvarez is horrified, his face ever so briefly awash with a wave of emotion. But still he remains calm. He does not whine or beg or weep. Some of them do. But he remains cool as the gun pushes against his head.

This is one thing O'Shay likes about his business. It trains people to a sound work ethic, it culls out those who are lazy or stupid. Or too greedy. The world, he thinks, has grown careless and fat. The wages are often too high and the work too easy. The price of failure is often no price at all. He sees it all the time in the police, their sloppy work, crude techniques, nine-to-five schedules. He despises a lot of police not because they arrest people and destroy people but because they almost always fail to appreciate excellence. They lack drive. They tolerate stupidity and do not weed out the stupid. As his industry grows, the authorities keep hiring more agents, keep raising their pay. And they keep failing. And this failure simply breeds the demand for yet more people and more offices. Waste. He cannot abide waste.

In his business, performance is everything. Just look at Alvarez on the carpet, a man behind in his payments, a man who has lost loads, and yet a man who comes to this city to do a deal, a man who expects no way out except one created by his own work and ingenuity. It is a wonder that such men ever go down to the police. But that is the cruelty of the business.

The police can fail and fail and fail and still their checks come, their jobs remain. Their work ethic has been steadily eroded by this pampering. Many of them retreat to offices and files and paperwork. They have lost the edge of violence, an essential for any success in the business. O'Shay has been in too many firefights. Sometimes he thinks the only really incompetent people he deals with in his business are the minions of the government.

O'Shay thinks that this will work, that this man can help him in his business.

They shackle Alvarez and take him to O'Shay's office. He hardly speaks, his black skin remains cool, his face concerned but still under control.

He says nothing about the voodoo altar. He seems relaxed by the turn of events.

O'Shay asks him if he needs to use the restroom because he knows that often men in such an instance desperately need a toilet. But Alvarez is fine, and says so.

O'Shay offers, "I want you to know this is not personal. It is simply business."

And then he gets down to business. Alvarez has two choices. One is self-evident. The other is to become part of O'Shay's organization and to work on its goals and not his own goals or his colleagues' goals.

Alvarez has a family, he has restaurants in New York, he has his life. He has many things he can lose. Or keep. He details all these things to O'Shay, it is hardly the time to withhold information. He explains that he comes from a large family, that most of his brothers and sisters are professionals. He explains that one sister is like him and that they both work hard in the business. Alvarez talks calmly of this, and that is his offer: he and his sister can deliver whatever is desired. And, of course, in return, keep what they already have.

"I can give you hundreds of kilos of heroin," Alvarez offers. "I can give you Southeast Asian heroin. My entire life has been as a liaison between the Colombians and the Dominicans and the Puerto Ricans and the blacks. I know Italians."

This is the moment O'Shay always enjoys, the moment when all those hours walking his dark house and thinking and plotting suddenly bear fruit. He observes such moments with full attention, waiting for a surprise that never comes.

Alvarez is like all drug dealers, on top of the world one moment, O'Shay thinks, and facing death and ruin in the next second. O'Shay prides himself on not being surprised. And yet, he knows that this lack of surprise is slowly killing him, extinguishing the fires inside him that keep him alive.

He sits there by his altar facing Alvarez and on the surface he is completely alert, but inwardly he feels like a robot executing programmed commands. And he does not care, that is always the sensation, this not caring. His big fists rest on his lap, his voice is soft and yet crisp in those few moments when he speaks. His eyes do not flicker, his face never reveals the slightest desire. A robot. The light in the office is sterile, a neutral light that makes everything pale and empty and tasteless. This light slowly saps the will from anyone O'Shay brings here.

They reach an agreement, just as O'Shay knew they would. But more than an agreement, they cross some dark waters and reach firm land on the other side. Because O'Shay knows he is going to release Alvarez back into his world, let him go to the airport and return to his safe places. And he knows that when Alvarez reaches his home, and his previous life is back, he will still obey. Alvarez will be sitting in one of his cafés in New York, slowly sipping a cup of rich coffee, softly giving commands to his staff, and he will feel O'Shay's breath at his neck.

Cosima likes to watch others dangle. She savors Alvarez's life: he is thirty-eight, he has children, he has a black Colombian girlfriend, he has businesses in New York and down there in the islands and now he is a man who was talking calmly an hour ago when suddenly a gun filled his eyes, a man who felt death near, then sensed a reprieve, then considered a deal.

She later tells O'Shay, "I don't think we should do this. We got a good thing here without him and his heroin. I don't think we should bring him in."

But O'Shay knows what Alvarez will do, that he will all but claw his way into the deal.

O'Shay plans, he thinks things through, he knows. He moves in a world where everything is covered by a fine dust and cobwebs, a kind of archive where even the future feels like the past. He suffocates as he plays Alvarez on a string.

Alvarez says one more thing: "You were really, really good."

O'Shay hands him a phone and Alvarez calls a man named Garcia in Colombia.

"Hey, how's it going? I've got a real good connection here now. I need stuff. I'm going to give you my sister to handle it because I'm buried in other business."

Garcia says that will be fine.

O'Shay senses he is in, has crossed some invisible line that separates people like himself from the rarefied air of international heroin merchants. He can almost smell the air at the other end of the line, the green tropical leaves, cool tile floors, leather furniture, small cups of intensely rich coffee, picture this stranger named Garcia in some villa with a view of the hills, the Andes rising like gods behind him as he answers the phone, makes a few brief comments, and hangs up, staring out into the clear mountain air, and in O'Shay's mind it has to be clear, just as there can be no sound around the villa except for tropical birds raucous in the trees, it has to be this way, almost like a dream because in an offhand way, it has been a dream for him, to see if he can play with the big merchants and not simply hold his own, or cut a good deal, but absorb them, take them into his world and dominate them.

For O'Shay, it must be a pure thing, with a mountain villa, a phone on a fine wooden table, the cool red tile floors, the roof also red tiles with the daily showers splashing off it, and no whine of motor scooters, no barking dogs, no bad music blaring from cheap radios, no, a clean thing, this serene place where the game is simple, straight, and lethal.

Alvarez is almost relaxed now. The altar with burning candles by his side, O'Shay feels…closer but not close enough. He is in play, now he must not care, let it float, see what comes. But he cannot stop the feeling of desire, of breaking out of his routine into the mountain air where everything is washed by rain and green and clean. And clear.

It all seems too easy, too fast. But O'Shay knows that is the rhythm of the business. Endless delays, days waiting in motel rooms, quick phone calls followed by lonely hours. And then it all begins to move and happens very quickly.

THE WATER FEELS EARLY-MORNING, the glass that happens before the earth shakes off the night and begins to stir. Color touches this water but does not own it. There is a wash of blue, a faded blue, splotches of darkness that whisper the depths waiting below the surface. A bank of trees swirls in the background, shades of green, hints of rose, a turbulence of energy with barely the suggestion of form. To the right, an actual limb, branches, the dense mat of leaves reaching into the scene. It is summer, when time is rich and full. The sun will be hot, and beat down on the water. The boy is in the jon boat, each hand on an oar. He wears a green long-sleeved jersey, tan pants, and a large hat to fight off the rays that soon will come. His face turns out and stares, the mouth flat and expressionless, the eyes almost hooded and peering but giving away nothing except the knowledge that someone is watching.

The boy in the boat is alone. He has no doubt gotten up early, hoisted the jon boat on his back, walked a dirt track through the country woods to the lake in gray light, and launched in this elbow of quiet water. He is living alone, been left at this place for a spell. The painting says this. His face seems reluctant to speak, almost the face of a mute. His body is ready to row and yet relaxed, the limbs suggest not a particle of tension. He is facing the shore, his back to the big water beyond. Soon the oars will dip, bite the surface, and move out into more open waters. A line will be tossed, the little boat will float, and the sky will stream overhead as he leans back and waits for a tug on the baited hook.

Joey O'Shay stands back, looks, and listens to the jazz softly walking through his house. He picks up the brush, makes two yellow daubs on the surface, leaves, yes, leaves to show that time is frozen in the peace of the elbow of the big lake. He stares down at himself as a child, free and easy in the full stillness of the earth.

Or it is a different day for the child. He is down at the creek and there is honeysuckle blanketing the bank. The boy feels eyes burning into him and turns and sees a man hiding in the vines. And then he smells the Evil Creature, that odor, that dark, dank odor. And with that the man turns and walks away.

He knows the man meant him harm.

THERE HAVE BEEN A SERIES of killings that have muddled O'Shay's affairs. As the Colombian heroin deal lumbers along, the murders provide a list of tasks he tends to.

He finds Johnny Boy. He is small, twitching, the eyes dart constantly, the habit runs $400 a day. O'Shay knows he will not talk, not now, that he is too armored by defiance. O'Shay simply wants a sense of him. In his mind, he knows he wants to choke him to death. But that can wait.

He tells Johnny Boy that his best friend, Big Dude, has been murdered. This causes a faint ripple of emotion to pass through Johnny Boy, then he is contained again. Johnny Boy and Big Dude botched a deal, so their colleagues hunted down Big Dude and killed him. It is a small matter, a business detail. Johnny Boy is thirty going on fifty, has been part of the organization since he was eighteen, when he got out of prison.

O'Shay can sense Johnny Boy was a bitch in prison, he can see this fact almost in neon across his face. He weighed maybe 120, and he was fresh meat. He was raped and stabbed eleven times and hid in solitary. When he was returned to the prison population, he fashioned a plan. He would be mean and crazy and safe. Upon release, he got into speed and killing. He was useful to the organization because of his appetite for death. But gradually he lost touch with his talent, his hits became sloppy. And now they have become pathological, serial slaughter.

...I would drown him and hope if there is a god, he would take this mutilated boy and rock him gently. Because he is forever lost...there is no hope...no fair shake...he is the creation of our need to erase what we do to each other... I will see Johnny Boy again soon...I will tell him I'm sorry for what I am about to do to him and I will describe the phases of his demise to him and I will apologize to him for what this world has done to him. I doubt he will do anything but curse me. But I would rather be around him and learn of the horror. He is reality.

But of course, O'Shay will destroy him.

He tracks down Johnny Boy's colleague, a twenty-year-old seeking to rise on the flow of Johnny Boy's violence. He can smell the weakness. O'Shay feels sure, certain of his powers. He looks into the kid's eyes and signals him that his life is over, that there is only one path left, confession, and by that act, redemption.

In such a moment, and it is a feeling flowing through the room, a thing concrete and yet invisible, in such a moment, they always talk because they know finally and for probably the last time someone will listen.

O'Shay and the suspects soar like birds in this moment, rise up high in the sky and careen, wheel on the currents, dive and rise again, feel powerful and free. They can drop the disguise and finally be, and that sensation of being is such a relief they will endure any consequence once they have tasted it, go gaily to their deaths, happy to be unburdened. And O'Shay knows they will talk to him not because he can kill them with his bare hands, not because he is cold, intensely cold inside. They will talk because they know he is bad and so they will not be talking to a stranger.

O'Shay is always alone at such moments, quiet and barely leaning forward, his eyes focused and yet expectant of nothing but truth. He is empty of questions, prods, yes, now and then a prod to show he is present, but no questions, no loud voices, no tricks. Tricks would be an insult. They must know their life is over and then they can finally talk. And they do.

The boy looks up and squirms, then comes to rest. He opens his mouth and a poem falls out...

O'Shay listens, the pieces, the little details falling into place.

Johnny Boy and the boy will go down, be quickly forgotten. O'Shay will sit on his black couch near those French doors, sit there drinking past midnight, his head full of details of his big deal, his new connections, he will sit there going over every possible alternative, making sure he has missed nothing. He has a thread of good heroin now.

But briefly, as he raises the glass to his lips, perhaps for a second, just a second, Johnny Boy and his sidekick will flutter across his mind. And then leave and go for good into the darkness just outside the door.

———

IT IS THREE THIRTY A.M. O'Shay sleeps flat on the bed, his face to the tin ceiling he has installed, when strange cries snap him awake. Off sounds always catch his attention, sometimes simply leaves fluttering in the pecan trees. But this is no movement of leaves in the night. He hears Awrr, Awrr, and he cannot identify the sound. He gets his gun and goes out.

He stares into the beam of his flashlight and sees a cat trying to kill a crow. The cat is hesitant as the crow uses his beak like a pile driver. One wing hangs, torn. O'Shay kicks the cat like a football.

He turns his attention to the crow.

O'Shay gets behind the bird, gets a stick and starts tapping and the crow stands, wobbly at first, and then tries to strut. Slowly, he herds it to the open door of his weight room.

The crow seems alert but calm.

O'Shay goes into the house, gets water and some bread.

The crow spends the night.

O'Shay builds the bird a perch where it can look out at his garden. The crow starts to heal. When it sees him, it makes sounds that O'Shay realizes mean it desires food. And when he brings food, the bird hops up to him and will take it from his hand.

He puts food out in the yard so the bird will get some air and when he does, other crows come and the injured bird pretends to be perfectly fine while the other crows are around. If O'Shay comes out at such moments, the injured bird hops back into the weight room, almost as if it were ashamed of its companions.

He senses the free birds and the injured crow are all part of one flock, now separated by the violence of that night.

He builds perches at different heights in the yard.

Finally, he gets a cage to take the crow to the vet. O'Shay pays a hundred and fifty bucks to have its wing mended and to have it released in a wildlife refuge. On the papers, he fills out the bird's name as Joe the Crow.

He tells no one of Joe the Crow or spending money for his medical care.

The vet had told him that when the bird was finally healed and it was ready to fly again, it would most likely go back to its flock. O'Shay wondered how the bird could figure out such a trek and decided maybe it was the natural way that white trash gravitated to shithole bars.

He misses Joe the Crow, the way he would come up from the weight room and tap on the French doors and then O'Shay would fling food out at him. He almost seemed to dance at such times. He was a natural gangster.

He is in his back room, sitting on the couch, when he senses something is watching him and comes alert. He looks out and sees a crow looking at him. The bird takes off and flies away. A few moments later the bird is back and staring again.

He tosses some scraps out, and Joe flutters around and makes calls and does his dance.

After that, O'Shay feeds him every evening. Then Joe the Crow disappears and O'Shay figures he's out grifting somewhere, maybe looting trailer parks or doing some kind of deal. One day, he finds a crow dead in the alley and he's stricken. He gives it a burial under his Carolina jasmine.

And then one day he hears the tapping and it's Joe and O'Shay rustles around the kitchen, gets a pile of scraps, a box of cereal and tosses it out for him. He can see the rest of the flock waiting up in the trees. Joe does his dance and eats.

Then he goes away.

But he comes back.

HE FALLS BACK, looks out at the night and can taste an earlier part of himself, a hard and mean part.

Two a.m, yes, two a.m. and he has been drinking. Joey

O'Shay feels the street beneath the tires as he wheels his muscle car through the city. His hair hangs, the beard hides his face and he is cruising, flashing through the intersections, tearing a clean pathway through the black, the warm inviting black.

Two guys run the intersection and he feels the metal crumple on the machine. And then they are gone, vanished.

He climbs out, it is summer, the air feels close, a storm must be nearing. The fender and light are ruins but the wheel still spins free. When he looks at the paint he winces. O'Shay feels it rising in him, his body tightens, his muscles fuse. But he is alone in the silence of dead.

A car wheels up and a redneck leans out.

He says, "Hey, you looking for two guys with a busted front on their car? Well, I'll take you to them. They're parked in a fucking alley with their lights off two goddamn blocks from here."

O'Shay nods.

He gets in his car and follows. The bent metal sighs, the engine hums.

The guy pulls over, leans into O'Shay's window, says, "They're just around the corner, their car is facing the other way."

O'Shay says nothing, gets out, reaches behind his seat and finds what he wants.

He and the other man walk silently. He feels excitement glowing off the redneck, he can smell the edge in the air, the tingle of electricity or something embedded in the coming storm.

At the turn of the alley, O'Shay touches the other man on the elbow, signals him to stop. He motions that he wants the driver. The redneck nods and shifts over to approach the passenger.

They walk softly.

The driver has his window down, his arm out, relaxed, his hand resting on top of the door. He is smoking and slowly exhaling.

O'Shay swings down on the driver's hand.

Then he pulls him from the car, crushes his face with his fist and when he goes down, begins the kicking. His companion stomps the passenger.

Ah, that was then, long time ago, O'Shay thinks.

THINGS ARE GOING VERY WELL, in fact, O'Shay thinks things are going perfectly. He has pulled off the biggest deal of his career. The deal was intricate, it meant doing things at multiple locations, coordinating everything about the buy, the kind of deal that dazzles a city as knowledge of it slowly seeps out into the stale air of the clubs where cigarettes burn lazily in the full ashtrays.

He is in charge now, he has learned everything he could from Angel Sonrisa, and now he has moved on and up and is doing his own deals. He has surrounded himself with a crew and tonight this team has been flawless, clockwork, like a fucking factory. Not one detail missed, not a single surprise, everything done perfectly.

So now they gather like players after the big game. They are in a café, a nothing place, but still for O'Shay it feels right. He is on a kind of contact high. Pizza, bring on lots of pizza and beer, endless pitchers of beer.

There is laughter as the team goes over all the little moves and details. They are happy with themselves. The days and weeks of waiting, of setting everything up, the delays that always happen, the unexpected changes, the fuckups, this is all behind them and in the end, everything came together and went down perfectly.

O'Shay feels like some master of the city, like the person who makes things happen and who commands experience to bend to his will. He has been building to this moment for years, doing all the shit work, being the strong-arm guy, then moving up into the world of the deals that escalate through marijuana and pills and cocaine to the high ground of heroin and tonight he has just done the biggest heroin deal of his life, serious weight, lots of complications and it has all come out exactly as he planned.

He savors the bitter beer in his mouth. All around him people are laughing and celebrating. The slice of pizza, warm cheese dripping off the edge, the cold beer, lights too bright in the bar and yet just right, it has all come together, it has worked perfectly, patience, just as cheese requires mold and time, patience, the absence of desire, the willingness to go with the flow and then, when the moment is right, is perfect, to move like a fast freight and close the deal.

Then the call comes.

The voice says he is down, he is gone.

O'Shay cannot believe it.

Sonrisa got killed. Sonrisa who taught him everything but whose teachings, it is now clear, are not enough.

Nothing is enough.

There are things beyond a man's control and these things must be acknowledged. And eventually tasted, and swallowed. But never accepted, that is essential, never accept them.

O'Shay can smell a child's hair, hears a dog with bell-like tones on the trail of a rabbit by the creek, watches a dark form slide in the water, sees the stars wheel overhead, all this, and tastes complete emptiness.

He almost has a list forming in his mind as he puts down the beer and realizes he cannot tolerate the sight of the pizza.

GENE TIERNEY IS RICH and cold and does not understand the man who is back from the war. But O'Shay does and that is why he keeps watching and rewatching *The Razor's Edge,* the 1946 version with its rich sets, black-and-white images, and rooms that seem designed to house shadows.

The man back from the war walks out on the promise of his life.

He says to his girlfriend, "I want to do more with my life than just sell bonds. The dead look so terribly dead—when they're dead."

Yes, O'Shay thinks as the scenes flicker past.

So the man goes to Paris, the man goes to the coal mines, the man goes to India where the holy man tells him, "The road to happiness is as difficult as the sharp edge of a razor."

For a spell, he is with Sophie, a drunk seeking relief from the memory of the car wreck that killed her husband and child. The man thinks of marrying her but his rich and cold former girlfriend stops this possibility by guiding Sophie back into the embrace of the bottle. She takes to drugs in the bad quarters. She is found in the harbor with her throat cut.

The man tells his cold and rich former lover, "Sophie's dead."

She says, "Dead? Do they know who did it?"

The man says, "No—but I do."

Ah, O'Shay thinks, yes, that is the weight, the knowing.

Of course, he has trouble with the man finding wisdom in some distant country where people wear towels on their heads.

But that hardly matters to O'Shay as he watches the film late at night, over and over. He hates things and so he loves a movie that tells him the things he surrounds himself with, the things he sells, the things people crave do not matter. Just as the fact that O'Shay thrives off their cravings does not matter.

Gene Tierney, cold bitch that she is, does not matter. Sophie floating in the harbor does not matter. Justice does not matter. The fact that Gene Tierney winds up in real life spending a lot of time in the crazy place—how is that possible given her perfect face?—does not matter.

This is not a good world.

The movie for O'Shay captures what does matter.

The man at least knows the question. The man at least knows the answer is not in the bottle or the pipe.

Or the fucking bond market.

5

ALMA SPEAKS ENGLISH PERFECTLY and looks like her brother, Alvarez. Joey O'Shay takes one of his men to meet her. He tells Matteo, one of his crew, to listen. His hair runs dark, as does his beard, his skin olive, he's Cajun and black Irish and yet he looks like a clean-cut American kid.

Alma is a small woman with straight hair, dark skin, plain features, jeans and an overblouse and she starts cussing as soon as she gets into the car—the plane, the delays, this talkative bitch that sat next to her, and her fucking husband, yeah, him, stabbed the sonofabitch, ah, she rolls on and on and O'Shay thinks some wind has suddenly come up in his car.

He says, "What? You stabbed your fucking husband?"

"Yeah," she says, "I'm married to a goddamn Puerto Rican and I stabbed him twelve times."

"Did the police come?"

"Yeah, they came."

"Well, Jesus, did that happen before you came here?"

"Well, yeah."

O'Shay thinks, this bitch butchered someone and then got on a plane and came to meet me? He tells her that the motherfucker might be dead and they have to figure out how to handle any heat that comes down.

"Are you sure he is going to live?"

Alma looks over quizzically and says, "Well, that sonofabitch is alive, hell, we're divorced now."

Slowly, he untangles her rush of words and learns she knifed her husband a while back and this trip is the first time she's been out of state since the stabbing. He learns that to speak with her is to stand before this gale of energy and language. He decides to skip taking her to a public place and heads for his office. He can tell she is rough, that she could whip her brother's ass. Her body is trim and yet solid, the muscular body some women possess without doing a lick of exercise.

He is intrigued by her moves, by her storming off the plane, by her effort to put O'Shay off balance. He is keenly appreciative of talent and he can tell she has talent. She is on his field and yet within seconds she is trying to make him play by her rules. Very fine, indeed.

The moment she enters his office she sees the altar and says, "Oh, shit, you a voodoo man. I've never seen a white mother-fucker keep shit like this."

"I just keep shit like this," O'Shay offers, "because some of them keep shit on me."

"Oh, I don't like this kind of shit," Alma says.

She walks over, examines the candles and skulls, touches the glass bottles.

She says, "You must have a pomegranate, I will bring you one for this altar."

O'Shay has had enough of this and gets down to business. As she sits there with the glow of the candles playing across her black features, he listens and he realizes she has moved a shit-load of heroin and has very good connections. She can be used, he thinks. They smile at each other.

She sees a photograph of a beautiful woman with full lips and a pretty smile on his desk, a face that is dark.

"You like dark women?"

"I don't like white women."

She breaks up with laughter. And she never mentions it again. Beneath the torrent of words, the jabber, she is all business.

"Look," he says, " tell me about prices."

"But," she says, "if you are wanting to get to know a heroin source—"

O'Shay interrupts and says, "Stop right there. What you are going to say is I must pay his expenses to get here."

"That's right."

O'Shay sits back. He knows that only a desperate fucking Colombian would pay his own way to come to this city to peddle his shit.

There is an etiquette that must be observed. That is part of the rhythm of the business, a thing O'Shay likes, a thing that reminds him that he belongs to a culture, a set of values that must be heeded if one is to be respected.

Alma explains the connection to O'Shay. Garcia is a dangerous man, she notes, but more important he is connected to some really bad motherfuckers. He will kill you but they are harder. Now, she continues, Garcia has a problem. Some black bastards in the Boston area have ripped off a kilo and he wants them killed. Also, he has a problem in Miami. His customer there is a priest and the priest has also ripped him off of a load. The blacks he will take care of on his own, if he must. But the priest is a special problem, in part because whacking a priest can draw attention. Also, Garcia is a Colombian and so there is a hesitation in killing a father, not a refusal but a desire to avoid such an act if possible. The priest claimed to have been robbed of the load, but he was blatantly lying. The Colombians found this behavior insulting, as well they should, O'Shay thinks. The father was a man using the Catholic church as a shield, and he owed them $480,000. What kind of a priest steals?

O'Shay listens and nods. He sees an opening here. On the one hand, he thinks, they know they must kill the priest. On the other hand, there are these factors, matters of faith that they cannot even say out loud. O'Shay feels free and powerful. He has no such matters of faith to deal with. Killing the blacks means nothing. And killing the priest means nothing. He can be of help. He is out of the area. He can send in hitters and they can leave and there will be no trace even as substantial as the vapor trail off a jet.

In the business, these problems keep arising. Given the nature of the merchandise, one cannot go to court and insist on the enforcement of a contract. The only enforcement left is death. And given this obvious fact, O'Shay is irritated that greedy little motherfuckers keep thinking they can rip off loads and go on their merry way. This violates all reason, it suggests stupidity. The Colombians, he knows, are sound businessmen. They seldom try rip-offs, and they kill if they are robbed of merchandise. He finds this fact comforting because it indicates that the Colombians are a sensible people.

So the talk continues and Alma slowly, but without saying so directly, lets O'Shay know that her connections are much better than her brother's, and that she has these connections because she has worked very hard and moved a great deal of product. She smiles and laughs and without seeming to lift a finger undercuts her brother, and brings the deal close to her. O'Shay is quietly impressed by her moves. And her anger.

Her ex-husband, the one she stabbed, had a bitch on the side and that is where all her profits, she had discovered, had gone. She wished she'd killed the sonofabitch and O'Shay believes her, has no doubt she would kill and kill easily. He has a feeling for such things, a fellow feeling, and now he can sense that she has that place within her, the one he visits, the zone where killing can be done in absolute serenity. He stares into her bright black face and feels less alone.

Alma tells O'Shay, "You're crazy. I never seen a white guy like you."

And they both laugh.

The talk does not take long because most of it is done without words.

"I was raised up in a swamp," O'Shay offers.

"You're crazy. You ain't like what I thought you would be. We can do this."

"Well," he counters, "let's have some fun doing it."

And Alma laughs and laughs and laughs.

He says, "You get Garcia to the point where he will come here, you let him know I'll pay the way and you let him know you've got a trafficker here, and he will make a shitload of money."

She says, "Okay."

But food is the key. She'd gotten on the plane without time to eat and now she is hungry. And hunger touches an important part of her because she is also a chef. O'Shay listens. He gets up and makes her a sandwich of peanut butter and marmalade. He knows people from her background are fools for marmalade. She brightens at the first bite. They are closer as the sharp, sweet jam spreads through her mouth.

It is all in the details, he thinks, as he stares at the stoneware marmalade jar.

Alma flies out the next day.

The waiting resumes.

O'Shay has his connections. He has Alma and her brother working on the matter, making those calls into Colombia, setting up the premises for a meeting in the city. And he has a

name, a man named Garcia who has lost a kilo to blacks in the
northeast and been ripped off by a priest in Miami, a man who
does serious business and could use some help in these house-
keeping details. It is beginning to feel fine. He is assured when
his instruments perform properly. And Alma is cold, ice cold,
and that also comforts him. She is a sound tool.

O'Shay sits alone at night in his back room, sits on the black
couch with light from the kitchen brushing against the French
doors leading to the patio, and he feels cold, the serene coolness
when things are fitting together, the cold that pervades the steel
door of a safe when the tumblers one after another click into
place and signal that entry is permitted. With the contract
killings on the table, he should be able to negotiate a reduction
in the price per kilo. With the marmalade on Alma's tongue, he
also feels calmer and safer.

He remembers once when he was a rookie in the old city jail
and one day in court watched as a city magistrate sorted out the
drunks and whores with indifference. The men all reeked of
Dumpsters and urine. The judge was moving right along, filing
various souls to their fates and making quick little bangs of his
gavel, when suddenly a dark man stood before the bench fluttering his hands and making not a sound. He looked terrified and
acted like some berserk person. The judge eyed him and figured
he must be some kind of mute. But then he was told that two
people on the staff knew American sign language and neither
one could make head nor tail out of the man's flailing.

The cops had found him wandering the bus station and
could get nothing out of him—except that he resisted them
when they tried to take him away. And so he stood there, teeter-
ing on the edge of an abyss, all but shipped to some state mental
hospital where he could be observed and warehoused and then,
most likely, forgotten.

Suddenly an old wino stood up, a regular who'd been boozing the downtown streets for years.

"Judge, Judge," he shouted, and began making a ruckus to get attention.

The magistrate looked over at this annoyance breaking the monotony of his tired courtroom and was about to have his bailiffs throttle the man when something made him hesitate, something he couldn't quite finger. He had the old wino brought up to his bench. Something was off, he sensed, something in the old drunk's behavior had a spark beyond the usual delirium of a juicer.

"Your Honor," the drunk sputtered, "I know that language this man is speaking. I can communicate with him."

The magistrate looked down with a cold eye that said, You, a fucking wino, know how to translate this loony's hand-flutterings?

The drunk raced into this pause, started talking a blue streak about how he'd once been somebody, a professor of anthropology on the coast, and how he'd done a lot of research somewhere down in Mexico, lived with a tribe of Indians who for some reason were afflicted with deafness and had evolved their own form of sign language and this man in the courtroom was one of them, he was sure, and he could remember enough of the gestures to talk to him.

And by God, to O'Shay's amazement, the old drunk did just that and discovered the man was an Indian from that tribe, and was supposed to be met at the bus station by someone, and then the person was late for some reason and the cops came and hauled his ass away and he was terrified to be lost in a strange city with a strange people who could not understand what he was trying to say with his hands.

So it got sorted out.

The Indian was hustled off and his contact in the city found.

The old wino fell back into himself and returned to cold streets and the warm dreams of cheap bottles.

The judge resumed the sleepwalking rhythm of his work.

And O'Shay thought, You can never be absolutely sure of anyone and you can never be absolutely sure that your read on what you're seeing is true. People have compartments and within these compartments are other boxes and you can watch and pry and beat the shit out of them and listen, and still you can only know so much, only get just so much sense of them. You can never be certain.

That's the fucking rub for O'Shay. Even a dog can misread a scent.

He goes into the kitchen, pours a beer into a Mason jar, comes back, and sits on his black couch in the half-light. He's got to think this through, he's got to be calm, he's got to get into the zone. The place where time stops and becomes solid under his feet.

The phone rings, he recognizes the voice, a fucking Irish gangster he's dealing with on the side. He sips the cold beer from the jar and listens.

"Let's do this on Monday," he says softly, "because if you get in a hurry, you die."

The caller hangs up.

O'Shay takes the remote control, hits the satellite radio service, punches in an endless stream of meditation music. The white walls soften as "Shamanistic Healing" flashes on the television screen menu.

EVERYTHING HAPPENS FASTER when he is young because he wants to get everything over with. He is at that point in his life, his early thirties, when many men rush.

They hit the door and enter fast since they know the guy sleeps with a loaded shotgun by the bed. He's a veteran who came home with a habit. Naturally, he feeds the habit by selling.

As O'Shay barrels into the bedroom he can see the guy coming awake, his hand reaching for that shotgun, and O'Shay has his finger by the trigger of the gun, a barrage of rounds less than a second away and it is that cold place where words like *life and death* are somewhere else, where all those discussions people have about ethics are mute, no, now it is now, his finger near the trigger, the hand reaching out of the bed for the shotgun, and O'Shay is young and knows he is not going to die but wondering if in a fraction of a second he is going to kill.

The guy stops, drops his hand from the gun.

The moment becomes a detail in the flurry of work.

But he cannot get such moments out of his mind. There is always this split second between living and dying and most people live unaware of it. And he lives constantly, he thinks, in that split second. He is having a beer, he is eating lunch, he is talking to a woman, but always this split second is there.

He runs wild, he does dope deals, coke is a big thing, lots of money. It brings an edge to his city, a cowboy attitude, and so people in the life live faster and faster and reach for their guns easily. He is in a house, there are machine guns, lots of drugs. He wanders into a closet and the guy has a ground-to-air missile.

The world is shifting and feeding and growing. The money and muscle in the drug world is booming out of sight, unrecognized by most, just like those split seconds between life and death, and for O'Shay they are becoming all the seconds of the day.

There is slippage, O'Shay cannot admit this to himself for years. He takes in everything he does and everything he sees and applies it to some learning curve as he masters his business. He is focusing on being that tool and so he refrains from noticing this slippage in himself as he slides from the person he once was and becomes this person he senses he must always remain.

He masters firearms, he hones reflexes, he listens to his betters and absorbs the key lesson: come out of it alive. Everything else, every remnant of courtesy, fair play, and decency must be stripped, abandoned, especially in those opening seconds when control is on the table, hands move toward guns, status is still up in the air. In those seconds, survival is the only standard. Someone will be alive when it ends and if it is not O'Shay, then he is out of the game.

He sees people with him take slugs in the gut, a round in the head, crumple and vanish before his eyes. They are trained, they are stone cold, they seem as talented as he is. Still, it can happen. And if he gives up his split second, his reflex, if he pauses for even a fraction of a second before firing into someone, he, too, is all but certain to go down.

He feels himself becoming hard, and feels this hardness becoming meanness.

He also notices at times his addiction. He does more of the work than he must, more than his crew requires. He has an appetite for it, for the moves, the guns out, the action. It is not feeding some sense of inadequacy, it is not proving something to others, it is not a buried appetite for sadism, he considers all of these things and moves on. He knows the lash at his back is something else. An anger, one he cannot share with others, or at least cannot speak to them about, an anger he can barely acknowledge himself, but one that stays with him all the hours of the day.

He feels a loyalty to those around him, the people he does the work with, but this is in some ways a mask for what he is really doing, a sense of brotherhood that cloaks his drive. He can smell at times the hair of a young child and then he will expunge this scent and move and his moves are hard.

He meets a man on a motorcycle, a man wearing colors and with many connections in the business. O'Shay slowly learns those connections. He moves into a shabby apartment with another guy, a blue-eyed blond boy who wanted to make his way in the life. The biker, an older guy, has his apartment in the same sprawl. O'Shay learns how to heat-seal drug packets with a phone book and a match, to test with Purex, the full bond test for cocaine, how to cut, cook, and package. O'Shay drifts into buying stolen cars, especially sports cars—one he even gets off a junkie for eighty bucks. Two guys come by one day and get high and start regaling O'Shay with how they beat some Oriental guy to death with a baseball bat.

He feels almost home, in a place where everything he has learned is suddenly useful, where all his workouts suddenly have a function, and where all his appetites can be fed. He's living on adrenaline, whiskey, Bob Seeger, and Jimi Hendrix.

There is the marriage but it is dying, not simply dying, it has been murdered by life itself. Evening—ah, how does he tell this to himself, how does he explain this part?—at dusk, he drops by to see his two sons. Sometimes, when it is very late, he simply slips into the yard, climbs over the big wood fence, and leaves toys and money for them in the knothole of a big tree. The wife, well, she is a ghost to him, someone he supports but does not want to share a life with. The boys are casualties of his desires, of his need for the life he is now living. And of course, they must be kept secret from this life for their safety and for his own peace of mind.

O'Shay becomes a creature of the night. He sleeps by day, rises late. Around nine p.m., he begins to face his world. After a while, he cannot sleep at night and when once or twice some warp of schedule forces him into such an arrangement, he is up at all hours pacing through the darkness like a cat in a cage.

But nothing really works, nothing really kills this pain in him, things simply postpone the deep visits of the pain. No matter how many bitches he drags up to the place, no matter how long he fucks, nothing kills this feeling in him, and he'll be humping or dealing and it will flash across his mind, the smell of a child's hair, the sense that none of his weight lifting has made him strong enough, not near strong enough.

He prospers and all the while he knows he and his partner are spiraling out of control. But he never touches drugs, he hates drugs, he sees movies where people use drugs and he thinks, They must be dumb cocksuckers to want to use drugs because no one has to use this shit. He doesn't and he is up to his ass in drugs. He drinks, he hits, he buys, that is different, that is on his own terms and it makes him feel clean. Drugs take away the edge and the edge is all that can possibly matter in life.

He does, he notices, drink his share of Jack Daniel's and beer. And he keeps his workouts going somehow, he pumps iron, he runs, he sleeps by day, he hunts by night, he sees his sons, leaves the toys in the knothole and he never relaxes, not with a glass and not with a woman. He feels totally alive and he feels one of the living dead. He tells himself that despite his ways his children must know that he loves them. And he drinks hard. And plows through the nights. Everyone who comes to their place packs a gun, and they sit around and shoot up, sometimes O'Shay and his partner help them.

O'Shay and his young partner get a bus and start using it as a kind of lab and base and they're doing five-pound deals of pure meth, and there are bales of money, some deals for hand grenades. The bitches are always around and they see this action and soon every titty bar in the city knows of Joey O'Shay and his partner and his scale of operation, and within a year O'Shay is cutting deals with more than a hundred bikers, speed freaks, and Aryan Brothers, and in this mob he finds sixty or seventy serious people, the kind with rap sheets and tattoos and a history in the life. And the nights keep getting longer and the booze keeps seeping into their lives.

One night, a guy says, Hey, I'll show you my lab.

O'Shay's driving with 440 under the hood, white leather interior, and good wheels, a machine designed for a speed freak. The lab is way out of the city in a barn, and O'Shay is cautious, everything seems black with night, there is a little house off a ways, and the dealer says, Oh, those are my tenants, don't worry. And it all goes smooth, and nice, and the engine rumbles as O'Shay drives back to his city.

But he realizes he is taking chances, making moves on impulse, making moves alone. O'Shay knows that in the blink of an eye he could be left out under the stars with a bullet through his skull and never even know exactly why he was murdered.

Business builds, one person leads to another. Their initial biker contact, well, sometimes he gets out of line, but the real beatings are saved for others. They move to another part of the city, rent some apartments for the work, and that's when O'Shay meets Bobbie, who is managing the property. The bikers get easier and easier, decked out in their colors and muscled up, but all hard juicers and often as not dealing and buying and stoking their bitches with the drugs.

He seems tireless and yet he feels himself slipping away from life. His partner can't keep up the pace, the drinking, the women, and he fades. So he takes on a veteran cop, J.R., tough as nails, trustworthy. But O'Shay and J.R. keep rolling deeper and deeper into this world that finally promises him two things: a way up and out, and a way to feed the anger within him.

He becomes a technician, or better yet, a student, and each deal is a way to sharpen his skills and extend his expertise. He is in a buy and in the closet with the seller where he stores his drugs and O'Shay has an ice pick in his sock and has planned a hypothetical escape route should things go bad.

Later, after the deal goes down peacefully and there is no problem, he checks out the escape route he had sketched in his head and finds that should he have vaulted from the window of the condo's second floor, as he planned, his fall would have been not two stories but close to four, because the entrance to an underground garage is gouged deep into the earth beneath that window.

Day by day and night by night, these little lessons come home, these bits of information that tell him how to read a person, how to anticipate a move, how to stay several steps ahead. He becomes almost a method actor in his ways of simulating a pose. He learns to watch a person's eyes to catch the first telegraphic signal of when, or if, they are going to make a move on him.

He learns the art of the deal and there are no business programs that help him in this study.

O'Shay has always seen too much and felt too much and this condition has made him lonely. As a boy, he began to shield himself from this bombardment in the air, in colors, in sound, in the movement he caught in the eyes of others. You take in what is at a glance, you never are allowed the lie of saying "I didn't know," you smell fear and love, you want to rub colors against you and eat the earth, the birds overhead are always in your heart, and the gas bubbling up from the creek bottom seems like your own subterranean life.

You learn, and you never quite remember the moment when this happened, but you learn to keep certain things to yourself, the stars wheeling overhead, the scent off a woman, the sheer delight in a flower, the love pouring into you when your dog nuzzles against your neck, the promise of the first real day of spring, the death in your nostrils when you feel autumn succumbing to winter. And the need for silence which you fill with the music of your mind.

There was a dog when O'Shay was young. He thinks of the dog now. He is in these biker kitchens, greasy skillets on the stove from some meal two days past, the counter lined with empties, the ashtrays always overflowing, everyone stancing, trying to take up space, talking trash and ordering the women around, a beer in the hand, then a swallow of whiskey, yes, standing there and staying alert for the deal and what O'Shay hears in his head is a beagle out there in the woods ahead of him and he can tell by the sound, a long bell-like voice, more the tone of a coon-hound than a rabbit dog, he can tell the beagle is on the trail and joyous with the scent pouring into her nostrils, winter is in the air, there must be a cottontail in the creek bottom, days and after-noons and dusk coming down, floods of scents with memories, the dog chasing game in her dreams, all this when O'Shay is a boy, that lonely time, the very worst, before he could escape into the world and find like kind, but hell, the dog dies, Dolly Mac-Groo, and decades later he stands in the kitchen, the air choked with cigarette smoke, beer warming in his hand, the deal, gotta focus on the deal, and watch their eyes and don't let one of their bitches get behind him, a dog with that odd bell-like tone, a rab-bit running by the creek, soon the light that says winter will lock down. There was a dog, he was young, and that was his last dog because he was lonely and the dog died on him.

That was and is his life, the scent pouring into his nostrils, the ears opened wide for the sounds of life, breeze rustling dead leaves on the forest floor, heart pumping and then at night stars wheeling overhead, wheeling right now as he stands in the dirty kitchen and listens and glowers because the deal is going down and his life is now the deal. This last fact he does not explain to himself, he simply lives it.

There is a place few people ever reach and this place is not a life obsessed by work, it is a place where work finally provides life. Not a living, not status, not a reputation but life, that same feeling a dog has running a rabbit by the creek in the dun colors of deep autumn. And escape, flight from loneliness, from fail-ure, from uncertainty. The gift of the deal is that it dominates sensation since failure is not simply undesirable but dangerous. The rooms, the stale air, the smell of failed cooking in the walls, the mattresses on the floors, the women fading almost as soon as they flower, the titty clubs, lap dancing, endless beers, sharp edges, and guns, all this focuses O'Shay and puts him on the path where nothing else exists.

He moves up at the same time. Through Bobbie, he latches onto a two-bedroom place in a high-rent building overlooking downtown and as he moves up he slides away from biker meth labs and into the new and lucrative world of cocaine. And heroin. He learns the ins and outs, how to cut, how many times he can step on it, how to weigh it out at twenty-five grams an ounce rather than twenty-eight grams as for other products. And the view is very fine from the eleventh floor.

He is wearing a vest, slacks, the beard is wild, hair an unruly mop. He stands in a private room of the bank where he can open his safety deposit box. The dealer is there with half a kilo of pure heroin. The money is on the table. Rising, rising into realms he sensed but had to find almost by following his nose. A metal box full of cash melting into a half kilo of almost pure heroin. The stench of ether from meth cooking leaves his clothes as he moves into finer things.

Now things begin to blur for him, the pace quickens, the bad nights in bad trailers are behind him, the deals and money flow. It is night, he dresses, moves into the city. There are meetings, conversations over weight and price, phone calls with coded language, the car rolls down the street, the black hours something to exhaust so that he does not have to face his inability to sleep, face his absence from his children, better to drink, hit those calls, make those meetings, haggle over ounces and prices, meet people and extend his reach like a vine creeping and yet growing and each night becoming more massive and penetrating new rooms and faces and names in the city, this other city that does not officially exist, that flows ceaselessly out of sight and mind of most citizens, the other city that provides the easy women, the blow jobs, the backroom games, the spike in the arm, the line on the coffee table in condos rising high over the lights of night, the city of pasties and G-strings and lap dances and expensive whores and bathhouses and stolen goods and hot cars and fists and knives and bodies dumped on the damp leaves in thrumming woods, the city where muscle has a role and so does aggression and there are no rules but the pretense at rules and there is no security but the quickness of the eyes and the hands and the fingers.

Here O'Shay meets people just like himself, some from the old neighborhood, meets killers with easy laughter, dead-end people who have found, for at least a while, a way around the roadblocks of life. It flows together, he immerses himself, he feels totally alive and yet toxic from the booze and the stress and the doomed lives he fondles and plays with before he sends them on their way to ruin. He discovers he is very good at what he does, that he is willing to work harder, that he is quicker with his hands, better with his mind, colder with his heart, that he is something he never imagined but now relishes: a predator.

The people come and go, hundreds of people, they go to prison, they go to bad shootings in parking lots, they overdose, they catch knives in their guts, they break down and melt away into other and safer lives, they are wracked with diseases, bitter about a woman, savage over a man, they sit in shit bars for hours drinking and smoking and waiting for the deal, or for the flicker of weakness in others, for some clue to the next act in their lives. All this passes before his eyes and O'Shay thinks he is exempt, got a pass, will never succumb, will stay on top of things and keep rising. He tells himself this is not ego, this is simply fact, plain as the trash in the gutters of the street, open as the woman when she spreads her legs at two a.m. one more time, a simple biological law that he lives rather than controls, like his long runs at night, his ferocious assault on the weights, the straight razor in his sock, gun ready stuffed in his britches, that's it, he's living natural law while others live statute law.

And the law, all the laws, are a lie. Bobbie takes over a new tower, and O'Shay bags a nice place in the building. Bobbie invites him to parties and he meets this big gay guy built like a football player who is peddling Ecstasy, a new designer drug. O'Shay is keen on its virtues, no needles, no pipes, no nothing, just open your mouth and swallow. What could be more American than downing pills to change your mood, a habit learned by every child at his mother's knee. But what intrigues him and gets him alert to his new world is that his gay Ecstasy dealer is dating the district attorney in a neighboring county and what in the hell does law mean then?

He starts hitting the clubs, the big ones with rope lines, and he travels with a posse of good-looking women, dives down into the lower levels where the VIP rooms are, he's Joey O'Shay, got the threads, the wild eyes and if you need it or you sell it, he's your man. People he's seen only in magazines are suddenly sitting by him at the bar and they don't look like the scale of their photographs but in the smoky air seem like toads with giant blank eyes.

O'Shay tips big, makes connections, starts making buys. He loves the clubs and the club scene, the booze and the women. It is all interracial here, the colors of his childhood paint the faces. Pimps stand next to lawyers, people fuck in the bathrooms and do lines. It is the kind of America that appeals, a nation where the barriers are down and the appetites are out.

And it is good business. His connections keep multiplying into new groups in the city. He meets artists, actors, antique dealers, people of all stripes bound together by their habits. The newspapers catch on to the scene and start running stories of drugs and how they are changing the texture of the city and O'Shay feels that he is part of something new and powerful.

He has everything he needs. Except meaning. And he wonders if his need for meaning is his giant flaw. He can't really talk about this with the people around him, they live beyond the need for meaning, they find what they need in acts. The glass of whiskey on the bar, the cigarette with smoke curling off it in the ashtray, the hours sliding past in the soft darkness, the phone calls and meetings and little deals and crimes and moments of flesh and there is no space for this meaning business. Nor appetite.

O'Shay knows there is something wrong with him, he is succeeding and his success is not enough and this is not right. But he also knows he must face this fact, not often, but still he must face it. He has women but he can't seem to hold them, or they can't seem to hold him. There is nothing but nothing for him. He is living too fast, that is what he tells himself. His life is a blur even to him and yet he is in total control and orders others around. He keeps building his crew and his operation, he fills his days with ceaseless deals. He almost never fails until finally he is at a point where he can achieve anything he wills. And he is empty. He wants meaning and he thinks he must be on a search for something he cannot name.

He hungers for the swirl of stars that he watched as a boy and yearns for as a man, the feeling of being back in the jon boat floating in a cove on a moonless night and the Milky Way smears across his face. Knowing that feeling, tasting those nights, how can anyone be so certain that there can be no God, no creator, no meaning to the misery he finds and the misery he creates? Hell, he thinks, the two guys who went closest to the stars and gases and void, Newton and Einstein, both refused to deny the possibility of God.

He has little shards of memories but not whole memories. He realizes he cannot recall most of the people he deals with, they are like mists that melt away after he has done a deal. He is in a room with a white woman doing a heroin deal. But first she takes out her breast and injects into it while he stands there. And for a brief second, O'Shay thinks she is an actual human being.

And then they do the deal and later the memory of that needle sliding into her soft breast rises in his mind and no matter how hard he tries, he cannot remember her name or her face. She has ceased to exist except for a breast, a nipple, and a needle.

At one point he visits a psychologist. The man listens to him and then says, "Well, you're 80 percent of the way there because at least you know you're fucked up."

One of his connections takes a hard fall and the lawyer says to O'Shay, "Make sure she dresses nice for court." She shows up in stacked heels and a dress slit down to her waist, her huge breasts spilling out, and O'Shay realizes that for her this is nice, that she is doing the very best she knows how to make herself winning and presentable to the judge and jury.

A partner goes down and O'Shay attends the funeral with a girlfriend.

The mother comes up to him and asks, "Why did they kill my boy?"

He stands there dumbstruck.

O'Shay's girlfriend keeps repeating to the woman, "I don't know, I don't know, I don't know."

The two women whisper in Spanish, fall into some language beyond language, a place of intimacy and pain, and O'Shay feels as if he is violating their moment together simply by existing, by standing there and breathing, by being the male, the gender that brings death to the houses where women wail. He goes home, gets into a hot tub, sinks into the water and begins to sob and when the water has become cold, still he weeps and convulses with his pain. A pain he only knew once before, when a child died in his arms.

THAT SAME SKY IS BACK, the kind of sight that would terrify a sailor at sea, a sky of gray streaks and blue and red and emitting a light that seems barely brighter than that in a tomb. The bearded smug man is back, that little disgusting smile, his head wrapped in some white turban. But his body this time, to Cosima's delight, is that of a chicken with white feathers, little feeble wings, and two brown scrawny chicken feet supporting his stumpy bulk. The claws on the feet are a brilliant white. In the background, camels move across a flat and then in the distance soldiers, barely discerned, penetrate his world.

His head is about to be exploded by a hailstorm of missiles, each roaring down and painted red, white, and blue. Ah, Cosima loves her dumb, innocent, adopted country.

Where else could she live as she lives? And where else could she feel this sense of belonging to something elemental and decent?

THE ENGINE RUMBLES as the cheap apartments come into view.

Everything is primed, the white biker bitch inside has been briefed: Joey O'Shay will look alarmingly healthy because he never uses the stuff, he tells people he's only buying to keep his topless dancers working hard.

He walks up, the complex a mosaic of blacks, Mexicans, and poor whites.

He passes an overflowing Dumpster, and even though he is outside, the entire world at this moment smells flat and stale. He sees raggedy-ass kids playing in pools of water left by a summer rain. Heat has fallen on the city like a steel plate and all he hears are flies buzzing on the overflow of garbage.

Two knocks, the door opens and the woman, in her late thirties, is skinny, the skin sallow, almost yellow. She is deep into the chapters of her life.

"You must be Joe," she says, without the energy for a smile. Her eyes are green from something floating in them, the hair, dirty and brown and clumped.

He walks in, that ice pick in his back pocket, the gun in his crotch. The smell of dirty dishes slaps him in the face, the stagnant water in the sink stares back at him like a blinded eye. The rug is stained from forgotten spills and smudged with dirt.

Joey O'Shay knows how all the rooms look, including the rooms he will never see, how dirty and wrinkled clothes will be heaped in the bedrooms, how scum festers on the sink in the bathroom, how ashtrays will always be full, beer bottles stand like icons breathing out stale fumes, how silence will never be permitted by televisions and radios that have never been turned off, bras will hang on doorknobs, yellowed bras, panties with streaks will lie here and there like doilies, the refrigerator will hold no food, never holds food, and on the small kitchen table the box of last night's pizza waits with the lid open and one last oily, cold slice.

O'Shay glances over and sees a fat black guy passed out on a rose-colored sofa. He's nodding, no shirt, black slacks, no shoes and what passes for dreams coursing through his head.

There is no air conditioner and a steam heat hangs in the apartment. Sewage backup wanders down the hall from the bathroom, a sweet rank odor licking everything in the place.

The woman disappears into a bedroom, then returns with her gram.

"It's real good," she says in that flat voice, "real go-fast."

O'Shay gives her five crumpled twenties.

She flattens them out one by one in her hand and then manages a weak smile.

She looks down at the bills and suddenly asks, "You been shooting dice?"

O'Shay smiles, pockets the gram, but says nothing.

The air grows heavier, the heat bears down, and all the odors of the small apartment slowly fuse and prowl like a beast.

The woman nods toward the passed-out black guy and offers, "My old man can get you heroin if you want some."

O'Shay shrugs.

He remains tense, alert, and yet relaxed.

Maybe she was once a titty dancer before the shit raked all the meat off her body. Now, well, who knows what she does for her nickels and dimes. Time is running out, she won't have anything to sell in a while and then the fat bastard asleep on the rose couch will move on and leave her to the stale air. Her teeth will go, there will be parking-lot blow jobs for a little taste to maintain. She'll snitch off some people for a break or a dollar. But basically, it is over for her. This is the end and soon even the seasons will pass unnoticed.

Suddenly a tiny kid with an unruly brown Afro floats from a darkened bedroom. His eyes are asleep.

The boy is maybe four, with hazel eyes and the eyes are flat, almost dead. He's naked and moves slowly, as if movement could be a dangerous thing to do.

O'Shay takes him in, a ghost who has barely been on earth.

The woman catches O'Shay's eye, turns, sees her boy, and yells at him to get the hell out of there, get the fuck back where he belongs. You hear me?

He leaves as quietly as he entered, leaves as solemn as a bishop. He really does not so much leave as seem to fade, to surrender what tiny flickers of life exist in his four-year-old body and return to the world of shades.

O'Shay says nothing. He turns, walks past the slumbering black guy on the couch, and goes out the door into the steaming daylight of the city. The rain of the night before felt cleansing at the time but the waters did not wash this place.

Out in the lot, he finds his car with the guy he'd left in case something went bad in that apartment. A Mexican junkie leans in the window trying to make time with a woman they brought along.

O'Shay gets in the passenger side and turns the blower on high, lets the air roar in an attempt to clean off the feel of the apartment, the green coating of the scrawny woman's eyes, the stench coming from backed-up sewage, the emptiness rising off the passed-out black guy on the couch.

He barks, "Let's get out of this shithole."

Later he comes back. Buys more. Gets some of that heroin from her old man.

She tells O'Shay, "My old man is cool."

It does not sound like love but it sounds like something, and in that apartment, something feels like a lot.

The odor is always bad, the dishes never get washed, the stagnant water is never drained.

The boy, O'Shay never sees the boy again. Sometimes he thinks of him but then he pushes the thought aside. The way the boy floated in and then faded away, that uncombed hair, no, the boy is too close to a ghost for O'Shay.

He does not want to go there.

HE DREAMS THAT LOVE WILL SAVE HIM. And then he puts that thought aside since he cannot afford dreams but must attend to deals. He lives, as always, a lie. He has a fine apartment in a tower, he buys and breaks people and casts them into the dungeons of the authorities. He drives powerful machines, cuts a swath through the clubs, has a look that is a mask. Still, this dream flickers in his mind in those moments when he forgets himself and what he is.

Sometimes she smells like vanilla and mint, sometimes she smells like a flower. The dark hair falls long to the waist, the full lips caress white straight teeth. They meet now and then for quiet lunches. He can sense an inner sadness in her. He savors her dark skin. She is just back from Jamaica, and wears a white sundress and gold hoop earrings. Her skin is coffee and cream as she glows and eats yogurt and cookies. This time O'Shay smells mint coming off her.

They cross the street to the park, fall lingers with gold and red leaves dappling the grass. A stone bridge arches over the ravine and below a creek purrs and is lined with hardwoods, honeysuckle, rampant vines and cedars and this green smear becomes huge pecans, oaks, maples, and sweet gums on the higher ground. Yellow leaves float by on the stream, she turns and her dark skin glimmers, her eyes become golden.

He wants to tell her he is falling in love.

She looks down at the water and tells him they should not see each other again.

He says he understands, he has caught the look others have in their eyes when they enter a café together. He says he wishes she could overlook his white skin.

He feels her lips on his cheek and then she says, she worries he could not accept her color.

O'Shay is stunned. He presses her body to his and she feels very fit and yet this moment suddenly very soft. And all he smells is mint.

They walk silently to his car, he drives without a word to the tower where he lives. They enter the apartment and undress slowly and without words. In the morning, she showers. As he drives her back to her car, he says, "I love you."

She puts her hand over his mouth and begins crying.

Then she gets out and he sees her walking to her car and she becomes for him a white flower disappearing from his reach.

He leaps out, bolts past other people, catches up with her and says, "Please, talk to me."

She says that the afternoon in the park, the night in the apartment were too beautiful to last, that people and the world itself will crush those moments and that they must accept those moments as all there will be for them.

He begins to weep, he begs. He tells her he will leave his undercover life.

She drives away.

O'Shay stands there thinking that a person only gets a rare chance at life like this one.

He swears to himself that he will somehow make her believe a life is possible for both of them.

And then he gets into his car, and he turns on the ignition and he is Joey O'Shay again and goes back to his world and his studied betrayals.

THE DEALS AND THE APPETITE for violence grow and there is a point when Joey O'Shay knows this cannot continue. But he needs a way out of the life he has created. He finds a book by a man named Viktor Frankl. The man is a Jew and O'Shay grew up with no more knowledge of Jews than of Martians. But he reads the book, which, along with Anne Frank's diary, is the first report after the war about the camps. The book is called *Man's Search for Meaning* and it explains how to make pain into something useful and good.

O'Shay is spellbound.

Frankl looks out and this is what he sees: dawn, barbed-wire fences, the towers where men watch, big lights that hunt in the night. The dawn seeps gray, and does not forgive. Men plod along in columns. He hears shouts. And whistles. It is all there on pages he writes later.

The man sees people hanging in his imagination, and the dead float and twist in his mind. He is past fear and into a place he knows only from his reading. He has entered horror.

O'SHAY STARES DOWN at this man and watches him shudder on the page.

He is like nothing O'Shay has ever found before in books. The man is a mirror and O'Shay sees his own face.

Trains mean comfort. Sitting in the seat or perhaps going to a club car as the world streams past and becomes a tapestry without scent or sweat or hot or cold. That is the joy of trains. Some claim trains in dreams mean death but this is a theory for those who think about dreams. For almost everyone else trains mean ease of movement and a comforting clickety-clack to register the progress of the journey.

For O'Shay trains mean freights, bums in switching yards, bad bars in warehouse districts. But now as he stares down at the man, the trains have a different life. The people are in the station, there are shouts, lines. The voices are thin and yet harsh and empty of love. The man writes, "Their sound was almost like the last cry of a victim, and yet there was difference. It had a rasping hoarseness, as if it came from the throat of a man who had to keep shouting like that, a man who was being murdered again and again."

O'Shay cannot shout. But he feels that strangulation in his throat, the weariness that comes from constant screaming. He realizes his screams are silent but screams all the same.

The doors of the train open, men with shaved heads enter. The man looks at the intruders, and notices they seem well fed. He grasps at hope.

O'Shay quickens at this thought: being in the abyss and yet seeking signs of hope. The place holds a thousand, fifteen hundred, in space for only two hundred. There is never enough food. Some men swap things for alcohol. O'Shay notes this, he understands the thirst. The man writes: "Under such conditions, who could blame them for trying to dope themselves?"

Yes. Who could? Who could possibly know the night and not need the drink, the drug?

Especially once the system makes its face clear. Those who

work in the gas chambers and crematoriums eventually are fed into the same mechanism and become yet more fodder for the system.

There is no hope.

There is no justice.

Yet the man persists.

O'Shay notes how the man persists.

He keeps reading. The pages should be too strange and distant for O'Shay. It is always cold on the pages and yet O'Shay has lived his life where it is warm. The man writing is one of them and O'Shay has hardly ever met one of them. O'Shay lives where people are brown or black or white (but an off-white), the kind called white trash. The man writing is cultivated, educated, a creature of learning and fine dinners and tasteful rooms. The man writing knows nothing of the streets, or at least knew nothing until very recently.

But O'Shay senses he can learn from the man.

He has found a guy who has been there, and O'Shay cannot even say what he means by "there." There is a line, and some are sent to the left, some to the right. The man is sent to the right. Later that day, he asks someone what this means, what happens to those who shuffle off to the left.

"Was your friend sent to the left side?"

"Yes."

"Then you can see him there."

The man looks up at a chimney shooting flame and ash into the gray sky.

It takes awhile for the man to understand.

After this book, O'Shay decides he cannot really talk to anyone who has not read the pages. He spends more time with the man on the page than he does with the people around him. He finds hope as people on the page die.

He tries to recognize parts of himself in Frankl's account.

They are such very different men with such very different passions and yet to O'Shay they are brothers.

But the pages make no sense unless you already know the pages. The man on the page went through his experiences before O'Shay was even born. And O'Shay had killed before he ever saw the pages about killing. Yet they are as fresh as the feel of the deal he finished the night before. They are immediate, like the ice pick he carries.

He can only really talk to the man on the page and he has never talked to him.

But he knows what he is saying. He knows you must lose everything, have all the things you value and trust stripped from you, be melted down in some vast oven, before you begin to understand the possibilities.

He stares down at the man who is telling others in his camp that he has a manuscript in his coat pocket, the work of a lifetime, a major intellectual achievement and that regardless of what happens, this manuscript must be preserved.

One of his companions looks at him and then smiles and says, "Shit!"

Frankl finally understands.

He erases everything that to that moment had been his life.

He stands shorn of his past.

He stands naked, stripped of illusions.

O'Shay says to himself, Yes, this man is my brother.

They laugh in the shower as they stand there with shaved heads and emaciated bodies. They are very thin and the water is cold, always cold. Still they laugh because it is a way to survive. If you laugh, you must still be in some sense human.

O'Shay nods in agreement.

Frankl records the need for curiosity. O'Shay once again is in agreement. He has made himself into a virtual chemist. He has forged his body into a weapon, he has learned the technical points of killing, the ins and out of laws, studied body language and dress until he can make someone in a glance. O'Shay devours everything that comes within his reach and cannot comprehend people who do not.

Frankl notes that with the coming of autumn, he is curious what the consequence will be for him when he stands outside in the chill still wet from his cold shower. He becomes clinical as the days pass, the temperature drops, and his body is subjected to this new blow. He learns he never catches a cold and this new fact feeds him in a way that surprises him. Curiosity, he decides, is part of being alive.

HE TRIES TO WATCH FOOTBALL on television on Sunday afternoon. The phone rings sixty-six times and each time is a hang up. O'Shay gets his AK-47 out and sits on the bed with the gun, watching the two teams play.

When he and his partner Red were younger, just really starting out, Red said, "There are people who know, people who don't know but are looking, people who don't want to know. But you know and you know who knows and you know who doesn't know. And that is all there is to knowing."

The phone rings again, another hang up. The teams scurry like little fat rats on the screen. The gun feels cool and comforting. The white walls of the house seem to shout.

The sound is off, the men collide and tumble in silence. O'Shay listens for something that may be coming at him.

O'Shay does not know. And that is all there is to knowing in this moment.

Years before, when he was studying the business and planning his ascent, he was sprawled in the dirt of a vacant lot with Sonrisa, both of them as backup, clutching shotguns, waiting in filth for a deal to go down. O'Shay tried to stay calm and let himself float, yet remain alert and ready to act instantly. Time became flat and solid and without beginning and without end.

Night falls and O'Shay wraps himself in it like a cloak. Streetlights cast beams into the vacant lot and glimmers rise off the broken glass and beer cans. Dog shit hangs in the air as heavy as incense.

He looks over and in the glow of the streetlight he sees a fat fly, a big goddamn blue and black and green fucker, just sitting there, still. He reaches over, and finds it is paralyzed by the cold. His eyes stay on the house, the place where the deal goes down, the place where he is backup should someone try to cheat or steal or kill, he keeps that house in sight, he keeps his hand on his shotgun and at the same time, the fly seems to speak to him, all the blue and black and green and fat ugliness, it seems a message, a signal, something, he can't name it, but he can feel it. Suddenly it flies up, O'Shay grabs it. Sonrisa says, "That's an omen." O'Shay finds an empty screw-top wine bottle and puts the fly in it.

Everything worked out. The deal went smoothly. He never had to storm out of the filthy lot, gun barking. Went smooth as silk.

But O'Shay kept the fly. It is still in the bottle. That was more than twenty years ago, but still, he hangs on to the fly and for Sonrisa and O'Shay the fly never dies, but simply stays drunk. Almost no one knows of the fly, and if they do and ask, O'Shay smiles and says something like it has good mojo. Or shrugs. He is not sure exactly why he keeps it, well, he is sure, it brings good luck, and sometimes before a deal finally goes down he brings it out of the safe and has his people touch the bottle before they go and close the deal. But he is never sure of luck, not even as a concept, he is simply addicted to its existence. O'Shay thinks with faith, he could open that bottle and the bug after all these years would fly away like a free spirit.

He's doing a deal and the people working with him cannot remember their fake identities. Or he's doing a deal and the people working with him offer information, say things they do not have to say, prattle away. Or he's doing a deal and hell, no one wants to work nights, or they don't want to work weekends. Or they are told simply to be silent and in the middle of a negotiation they start talking and make the drug guys he's haggling with suddenly raise an eyebrow. And then there is the problem of agencies cluster-fucking and fighting over who gets the credit. Sometimes he drives a deal to a terrain so rich in possible sources and money that the agencies simply want to end right there, bag their headlines for a bust and call it a day. O'Shay is in a war and for the most part he seems to deal with half-hearted warriors who earn too much and do too little. And lack the hunger that is essential for the kill.

But there is this element, as real as a steel beam, and he cannot control this element and so he needs luck. The bullets that have passed near his head missed him for no reason that he can comprehend. A blue, green, black fly, fat and resting in a bottle. It is necessary.

He keeps it in a safe in his office right next to his voodoo altar. And when he is about to close on a big deal, when he has figured out every possible eventuality, he takes that bottle out and holds it in his hand and looks at the fly. And waits for a signal.

The phone rings again. A hang up.

The game goes on. In silence.

ARCIA FINALLY COMES from Colombia. First he will take care of some matters in Miami, then come to meet O'Shay, who has paid for his ticket. Alma, the dark, marmalade-loving Dominican, flies in from the East Coast to assist at the meeting. She tells O'Shay that Garcia is a fair-skinned man with gray eyes who will sit there politely without a word but his eyes will take in everything. O'Shay thinks about this.

He sends a tank-sized vehicle to the airport. Alma will go into the airport to get Garcia. Garcia and O'Shay will sit in the backseat, speak in Spanish, do business. The driver has been out on an endless deal and has only come out of loyalty to help O'Shay push the deal even further. O'Shay tells him: You will not say a word.

Then O'Shay tells himself he doesn't give a rat's ass what happens. This is part of his ritual, his way of getting his mind in the right place for the deal. Still, he worries about the driver, a muscle-bound guy who no matter how much O'Shay disciplines him is always liable to open his mouth. O'Shay has already drilled Alma on his business, on how he does not travel, how he simply ships things by plane and truck and how his people handle the shipments. And how they have been taught to keep their mouths shut about his planes and airports and routes and methods.

Garcia doesn't look too good to O'Shay, who wonders if his queasiness is from the flight delay or some kind of indigestion or perhaps some deeper register of caution and worry. He has hardly settled in the backseat and Alma is translating when O'Shay's Spanish falters and suddenly the driver pipes up, too. He tells Garcia they'll take him out to eat, show him a good time, all this shit O'Shay has planned but does not want announced, hell, O'Shay, sitting in the backseat with the Colombian, wants nothing announced. O'Shay's still getting a sense of the guy, smelling him, taking in his moves, his polite and elaborate Spanish, his good manners and cool nature, filing all this and trying to get a feel when his driver starts this babble and although O'Shay cuts him off, he can tell there is more on the way.

Garcia is fair-skinned, almost German looking, and his eyes are not gray as Alma said, no, O'Shay thinks, they are violet, almost unearthly. His slacks are Italian, as are his loafers, his watch thin, his two rings small and tasteful. He is well built, maybe thirty-eight, around five seven, maybe 165 pounds, and he seems relaxed, poised inside his soft, almost silken shirt. His shoulders are broad but his hands are small and manicured and O'Shay can tell at a glance that they have never done serious labor. His hair is short, well groomed, his face shaven, and he seems so pale that O'Shay again wonders for a moment if he might be ill. A faint fragrance of cologne floats off him as he settles into his seat.

They talk casually in the car. Garcia speaks of his family in Colombia, of his daughters and how they are the loves of his life. And he speaks of his other woman in Miami. O'Shay is surprised at his openness, at allowing facts to be known that might be a tool against him.

And yet O'Shay notices also that Garcia takes everything in, stares at people, stares at the driver, at Alma in the front seat, stares at people outside the car window, and does none of this furtively, simply does not care if he is noticed in his actions.

They get to the hotel where O'Shay has gotten Garcia a suite complete with a hot tub, and the driver and Alma take him up to his room.

Garcia says he would like to rest for a while.

But the driver keeps breaking in. He gets Alma to translate and he tells the Colombian, "Look, you did the right thing by coming here and meeting Joe. He's a transporter, he can move things for you. He has jets and every time he fires up the jet it costs him ten thousand dollars, a fleet of Gulfstreams..." and he rambles on, hungry to sell the deal. The Colombian stands there like ice, his face a blank, taking this in and wondering why this person is talking to him about serious business matters.

Later, when Alma tells O'Shay of this conversation, he thinks this deal is dying, that this breach of etiquette would destroy anyone's confidence in his organization. He will deal with the driver later. But for the moment, he tastes what he always hates, the intrusion of fools into his plans. He revisits one of the two problems of any drug deal: the endless waiting and the eruption of unexpected complications.

O'Shay picks up Garcia after he has rested and takes him to a nice steakhouse downtown on a bluff. He thinks maybe dinner will get this matter back in its proper space. And as they eat he notices that Garcia for all his control and calm cannot help himself when a good-looking woman walks past. O'Shay smells a weakness. They talk about the food, the view, the weather.

O'Shay says, "Let's go and have some after-dinner drinks."

Garcia says, "Certainly, but I am tired. It will have to be a short night."

So O'Shay backs off and takes him to his room. He can sense

an unease in Garcia, that some warning flag has gone up because of the driver's impertinence.

He remembers again a piece of advice he was taught by Sonrisa years ago, when he was still rising: "Never rush into a deal. If it doesn't feel right, back out. You can never be burned by a deal you do not make."

And O'Shay realizes that must be exactly what is going through Garcia's mind at that moment as he sits quietly in the calm of his three-room suite and stares out at the blackness of the night.

O'Shay knows he must do something to regain his standing with Garcia. He considers the appetite for women Garcia has displayed. He decides to take Garcia to a club the following night. He takes Jaime along because he will treat Garcia with the proper respect. After all, Jaime has been in the life for years, parties regularly with leading members of the Mexican organization, knows what to say and what to leave unsaid. And is not disconcerted by the thought of large amounts of money or large amounts of weight.

In the club, they watch the girls on the poles, they eat and drink and talk about everything but business. Garcia's eyes caress the women and lick them. The Colombian is in his serene space, the place where he allows himself fantasies. The women come by the table, gyrate, and O'Shay tips them $100 and $150 a crack. But he will not let them lap dance. As a courtesy, he lets Garcia start picking the dancers.

O'Shay and Jaime ignore them, and Garcia pretends to ignore them but there they are, a foot away, their scent wafting, and the women are happy because no one is trying to touch them, that is not part of the deal, not part of the manner of the deal, and so the three men sit and talk and pretend they are meeting in some tearoom. O'Shay thinks, I'm spending money out my ass and ignoring everything I am buying. But he knows better, he knows exactly what he is buying, an image, an indifference, an armor of cold blood that he can almost see reflected off the Colombian's eyes.

"What do you want to eat?" O'Shay asks his guest. "Do you want shrimp?"

The money keeps flowing. O'Shay raises his eyebrow and two men come from a side table and are introduced to Garcia as O'Shay's men, Matteo and Angelo. They are polite, deferential, and they retreat after their introduction. Everything is moving again, and moving smoothly, all those hours in the dark planning each nuance, all that time is paying off as things return once more to the format and tempo O'Shay had imagined. It is not about money, or heroin, or even pride, it is for O'Shay at this moment, as tits bounce in front of his face, about control, and he can feel that control returning to his hands.

Jaime leans over to Garcia and says in Spanish, "You know Joe doesn't really like coming out like this to these places, so I'm kind of surprised he brought you here."

Finally, O'Shay, who is drinking and marking time, waiting out the needs of his visitor, asks him if there is anything else he might care for. And Garcia says, yes, he would like to see women have sex with each other.

O'Shay has a private room arranged, and then he and Garcia pick out a blonde, two Latinas, and a tall, slender black girl. They all have huge breasts.

O'Shay says, "My friend here is from Colombia and I want you to be sure to please him," and the women all nod obediently.

Once in the room, Jaime assumes the role of director and suggests acts and positions. Garcia keeps drinking and glows with excitement. Round full breasts, manes of hair, tongues in the pussy, writhing and writhing, bending, one of them an acrobat, pulling down each other's G-strings, fingering, and Garcia is transfixed, he is gone to some other place, he is almost drugged by the scripted orgy passing before his eyes.

It is all back within reach, the error of O'Shay's subordinate can be forgotten, the climate of the deal can be restored, the ease that must exist before O'Shay and the Colombian can reach out for a transaction, leave the safety of their lies and fronts and for the briefest moment buy and sell in violation of law. The music thunders with fast disco beats, the women move, feign, sigh, suck, lick, arch their backs, lights dappling off their skin, only their eyes lying about their real desires.

Garcia asks, "Can you buy one of them for me?"

And O'Shay says, "I'll ask."

But then without warning the alcohol hits Garcia full force and he decides against a woman in his room that night.

When O'Shay drops him off, he tells Garcia, "Now don't drown in that hot tub. We have business to do with each other."

BACK WHEN HE WAS RUNNING with the wild boys, just starting his climb up and out, O'Shay lived in a small house in a bad neighborhood. The house is buried in a honeysuckle vine and sometimes when he smells it, he remembers swaying on that swing as a boy with his girl who smelled so clean and had those eyelashes like butterflies floating on her sweet face.

He discovers a tomcat lives in the vine, a beast with no tail, one ear, one eye, and huge scars all over his hide. O'Shay wonders if some dog had mauled him. The cat is wild and savage. This is good. O'Shay hates domestic cats, in part because they barely bury their shit, like the sorry-ass scum who live on the street.

The cat never seems to kill a bird but decimates the local mice and rats.

Once a woman comes over and sees the cat and says, "He looks evil."

O'Shay snaps back, "Leave that motherfucker alone. He's a survivor."

He starts leaving scraps of meat and water out for the cat.

At first the cat is leery and ignores these offerings.

Then O'Shay notices when he comes home that someone has been drinking and eating. In time, he can sit on his little porch and the cat will feed.

One morning O'Shay is up very early, he's never been good at sleep, and he is out on his porch with a cup of coffee.

A big tomcat comes over from next door.

The one-eyed, one-eared, no-tail cat from the honeysuckle vine tears into the tom and O'Shay sits there listening to the screaming as the cat rips his domestic neighbor to shreds and chases him down the street.

And he feels good and not so alone.

POPSICLE'S VOICE IS NOW URGENT. Not simply begging or pleading, not asking questions about why this is happening to her. No, now it is urgent. She can smell death and she can tell the smell is coming off her fresh young body.

Her thin legs and thin arms and small torso, all of these things will grow still, and the camera and microphone can almost capture the sound of life ebbing out of her. And then suddenly he will pull the bag off, her face will greedily suck in air, once, twice, three times even, her eyes have grown to the size of headlights and all these eyes can see is his face, the camera, and the face of the person behind the camera.

The questions hardly ever come now.

A few gasps, but not questions.

She has found a rhythm that takes her to the edge of death and then just when she thinks she cannot persist one more second, or even at moments when she has gone long past where she imagined she could still survive, then suddenly without a warning or a word, there will be a reprieve, the plastic pulled back, the sudden feel of air in the lungs in the stale chamber of the apartment.

The camera still has what seem like expanses of torpor when it does not move and then suddenly awakens and peers closer at her dying face. And the face is not peaceful, not slack, but tense, almost a brown mass of knots. She looks so young, so soft, so pure. She looks like someone who should be in a jumper in the school yard, not out on the streets of the city selling her body.

There is a sweetness to Popsicle.

She may not be well spoken.

But, still, there is this sweetness.

And the camera, whether deliberately or accidentally, records this sweetness.

O'SHAY RIDES BACK through the darkened city and he feels the possibility that the deal can be salvaged. He hates to lose and so keeping the matter in play gives him some satisfaction.

He knows this is not a businesslike way to think, but still it is essential for him, just as it is essential to go too far, to pass the borders of accepted behavior and venture upon dark waters where you must be on guard and yet relaxed, must be dangerous and yet calm. Success breeds boredom and boredom breeds inattention. And inattention will kill a person. He wants to know if he can operate at the highest levels of the business.

And so no matter what he tells himself, he knows that his ego is in play. He wants to make this work simply to see if he can do it.

He can talk with no one about this.

He can barely even let himself think it.

The money will not matter when he is dead, nor the cars or hot women. But being there in the game, being in the moment, that can at least keep him alive until he is dead. And that counts for something.

Why else, he thinks, is he in this play? He has clean and easy work with marijuana and cocaine. He can barely break a sweat doing such deals. He feels like a cardsharp playing with innocents.

The players simply aren't good enough. Or mean enough. O'Shay walks into the room on such deals and knows everything that will happen before it happens—he fights this sensation, fights confidence, fights anything that makes him less alert, but still the sensation hangs in the air—and he hates this feeling and thinks, Maybe this is what the dead feel, this lack of surprise, this staleness seeping into the pores.

He hates sport fishing for the same reason. It is too predictable and it is oddly enough too much like business, the hunter and the prey. Because in his business you are either a predator or you are taken down. This is not a place where the lawyers fix it for you. There is no safety net. No fallback. Just the fall.

He tries not to think about this, not the fall, or the edge, no, the tedium, the hook set, the wiggle of the line, the strike, and the surprised look in the eyes as the beast breaks the surface. That's why when he fishes it is the way he did as a boy, a trotline, bottom baits, check it in a few hours or in the morning, haul in the fools at the end of the line. The buzz of flies as you lean your back against the tree by a creek, let the hours flow, and know that under the water they come to their doom.

He needs the deal, the heroin, to maintain, to persist.

Without it, well, he does not even want to think of that possibility.

Thank God for alcohol, the one drug he can trust, the one drug that pulls him past his anger. Thank God, at the end of one of these fucking days, like this one with his driver talking out of turn to Garcia, he can sit alone with a glass and wash it away.

Boredom he's been spared in life, but anger, that comes.

That fucking turtle, that big snapper when he was a boy, he can still see the huge fucking thing gliding down the creek at dusk, hunting, offering bait, waiting but not waiting because that big fucker, that's how he got so big and old, it had to know, know they'd come to his waggling tongue, know his jaws would snap and tear into their flesh, had to know. And in that knowing, it couldn't have been bored, not possible, naw, it had to know and have been serene in that knowing. And he is still unfinished, not there, not yet.

O'Shay takes some more of his drink. Wonders if he wants to go on this way, wants to follow this thought, find out what he really means by stillness and aloneness. He can feel a bullet go by, smell the light scentless scent of a child's hair. He flinches and sips some more. He sees the snapper slip down the dark waters at dusk.

Soul? O'Shay thinks this is a word he can barely say aloud and now it is booming in his head. He plunges on, he stops resisting. He has lived so many deals and he has explained these deals to himself in different ways at different times. In the beginning, it was a career, a chance to see if he could rise. Then it became almost like a science, a whole new field that the business magazines and books hardly knew existed, a kind of exchange with no ticker tape and no commentators, a naked place floating in the night where he had to study and study in order to create and protect his portfolio. Finally he felt like a textbook, one never written and never taught.

THE MAN ENTERS A NEW WORLD and some things in this world exhilarate him. He can tolerate cold. He can sleep under impossible conditions. In the barracks, the men sleep in tiers on boards. Each tier is six and a half feet by eight and nine men share this space and two blankets. Shoes function as pillows and in winter the nights are bitter cold. Sleep comes easily and is a delight. Sleep erases the day.

O'Shay lives almost without sleep. Frankl finds that in the camp even formerly light sleepers pass the night in bliss.

The man also learns that despite the filth, the lack of toothbrushes, for example, his gums have never been in better shape.

Days go by without washing and yet the men do not get infections. People can get used to anything, can survive anything, the man decides.

He remembers a line from Dostoyevsky that underscores this ability to survive. But he does not understand how this ability to survive functions. And he thinks of suicide even as he notes the new health in his gums. But he also realizes that he no longer fears death because should death come to him, he will be spared the terrible temptation of suicide.

O'Shay does not feel alone. He has found a companion who learned the lessons that lash O'Shay and mastered these matters before O'Shay was even born.

HE GETS UP FROM HIS BLACK COUCH. Moving softly for a man of his size, he pads through the kitchen and glances at the stove where he once cooked but now cannot seem to find any interest. Past the little breakfast nook, the paneled doors on the cabinets glowing with stained glass, nicks a corner of the dining room with paintings booming surf on the wall and the long polished wooden table catching faint beams from the light by the sink, then into the living room. Here he stops and tries to find some other space for his mind. He glances at the white fireplace, sensing the hulk of the baby grand piano in the corner, feels with his hand the soft cushion of the white sofa just before him in the semidarkness, tastes the stale glow coming off the streetlight that seeps through the frosted and stained glass of the front window. He notes once again that he has a house that slams shut all glimpses of the city.

Suddenly he goes back. He was young then and hard and mean and he didn't so much talk to people as snap at them. And people were afraid to talk to him. So he went out into the woods and shut a door and opened a bottle and for three days and nights, he sat there alone with a pen and paper. He would write a line, then stare at it, then scratch out a word, put in another word, then change the verb, then stare some more, then draw a line through the sentence, almost tearing the paper with the force of his rejection. He'd pause, sip his drink, then try and write a sentence again, just one true sentence, one that did not lie.

He did this for three days, and he created something he has never looked at since. Words, fucking words, that sketched something he called evil, sketched something he sensed in himself but could not quite spell out.

What was he trying to separate himself from? The deaths? No, that would not be honest. And he was not going to lie about this fact.

But he needed to feel that he was different from the people he dealt with. And this difference was not about being intellectually superior, or about having more power or money. He did not even have to be more moral, or moral at all for that matter. He was reconciled to what he was. But he had to have a purpose, a purpose he sensed would set him apart and in some way, perhaps for a while, it would keep him alive.

He knew where the danger came from and that is why he went to the woods. The only person who could get to him was him. The others were too slow, too fixated on their immediate needs. The others were too easy.

So he came back from the woods with a fistful of pages and now he cannot even bear to read those pages. He can just smell a boy's hair.

He turns from the living room and enters a hall lined with paintings. The light is low, almost a seepage from his study, and he can barely make out what he seeks: on a field of green a man in an old-style baseball uniform is sliding into home plate.

O'Shay can just faintly pick up the green in this dim light but still he fills in all its intensity with his mind. He knows the grass, knows the man who is his grandfather, knows the need to slide safely home. After all, he painted it. And that is what bothers him. He cannot paint anymore and he is not sure why. He'll get ideas, he'll think about a composition. But he will not pick up his brush and paint. He looks down at the drink in his hand and gently swirls it, moves his weight from one foot to another and takes comfort in the soft creaking of his perfectly refinished wood floor. He leans toward the painting, almost sniffs the grass, then opens his ears to try and pick up some crowd noise from the sidelines that are off the canvas.

He returns to pacing, a lap into the kitchen, a brief moment in the back room where the couch waits by the French doors, then through the kitchen again into the breakfast nook, the dining room, the living room, down the hall. He loses track of time, that's good. He cannot sleep. That is bad. Especially bad because he has gone over the deal endless times and is not worried. They will sell, he will get his heroin. He could smell it coming off Garcia as he watched the women dive and lick in the back room, he could sense that Garcia came here to deal and it would take an enormous warning flag to stop him.

No, it is not the deal that is keeping O'Shay up. It is something within the painting, some lost thing, some toll he has ignored and now must pay. He opens a can of beer. When he finishes the last of six, it is 4:50 a.m. He has reviewed endless deals.

He can make the faces blur and seem simply like the products of an assembly line, all the same, all blank in expression. But if O'Shay relaxes for just an instant, his nature takes over and the faces focus, the eyes have color, the skin tone, the women scent, the men body language, and then the words come floating back, the long-vanished conversations and hagglings, the tension and the rush as the deal goes down. And the people have ages and families and the names of their children dance across O'Shay's mind, and their little weaknesses, their cravings for fine steaks, but not too rare, or fine cloth or certain songs or easy women or a day on a boat. If he relaxes they become all too human.

But if he does not pay attention, if he simply goes to some place where he feels at ease, he will die, and not from some interior release but from a bullet. Or a knife. He will take a proffered drink and sip it and suddenly collapse from some poison.

Once, he is in a hotel room, the deal is going down, and he gets up and paces to the window and sees a strange car wheel in. There is something off about the car, too flashy, out of place among the vehicles of this nondescript hotel's parking lot. A kid gets out, maybe late twenties, black hair, light skin, good clothes, and strides into the hotel with a purpose.

And O'Shay is standing there, the deal hanging over a coffee table at his back, the eyes of the other guys boring into him, and he knows something is off about the car and the kid, something not right for this minute at this hotel at this time, and he can do nothing about it. He cannot fail to notice and he cannot act. He cannot even break his train of thought with the deal and yet at the same time, he must entertain the thought that an elevator is slowly rising with the kid in it and the kid is delivering a kilo, a sample for the meeting, or the kid is some hitter dispatched to visit this very meeting. So he pauses but does not pause, does a gun check and yet never physically makes a move to touch his gun, drifts off from the bites of sound coming from the negotiating table and yet never leaves the thread of talk at that table. And feels alone.

His drivers, gunmen, his Cosimas and Bobbies and Jaimes, they are only good up to a certain point. All the help can do is kill the guy who kills you. When he looks out the hotel window and down into that parking lot and sees that car wheel in, he knows he is well past the point where anyone can help him but himself.

He glances back into the room, the vaulted ceiling, the cream walls and notices a frieze of leaves all colored rose and green and white and shimmering on a broad band that binds the room like the tape on a kilo of heroin. He keeps talking but he can smell the rank odor coming off the frieze of leaves, can taste that moment of sweetness leaves have after a brief dawn shower in spring.

When he has time later to have his people look into the matter, he learns the kid came to the desk and used the ID of a dead Colombian drug dealer and that he turned out to be the grandson of the dealer and came and left the hotel for no reason at all, or at least no reason that had anything to do with the business of Joey O'Shay.

But that is the important part, that things happen for no reason and yet in order to stay alive coincidence is a forbidden thought, a weakness that will make him miss something and he knows he only gets to miss one thing and it is over.

He has operated for decades and yet no one in his city knows who he is, save those who do business with him. He has no reputation on the street, only in the rooms where he negotiates. He leaves no trail, not a track. No one who sells him drugs can find his house, or reach out to him except by phone. And the phone numbers keep shifting. He wears no jewelry, his cars blend into parking lots. He never drops names. He never offers information. He is business and now the business is growing yet shrinking and it is shrinking because he knows the business so well that it seems to thrive in a very small space.

He should feel better and he knows it. He should feel more, at least feel more apprehension, he thinks. He should, yes, he should feel anger rather than this sadness that seems to fall on him like coastal rain, like one of those rains that comes in softly during the night, a rain without thunder or lightning, without wind, a rain that moves like a cat and then simply stops, stalls, and falls without drive or movement, falls hour after hour and day after day until everything is limp and moss grows on clocks and yet the rain continues without any hope of ending and with all memory of its beginning erased.

He can no longer understand the nights. Now they seem to stretch forever and yet seem only thirty minutes long.

IT BEGINS WITH ALMOST A HUM, like the thrumming of insects in the woods. They are at lunch, a good Mexican place with under-currents of the ocean, and Garcia loves to fish and so O'Shay feels a soft spot and tells of the lake, his boat, the line sailing out over the calm waters. He tells him of catfish pulled up from the bottom, and Garcia offers tales of sea bass taken off the Colombian coast.

They are rolling, the food arrives, the words are warm, two men being boys and having a fine meal. O'Shay has given the waiter $100 for a table outside on condition that he seat no one near them. Garcia watched this moment and O'Shay realized he understood English, whatever his denials.

And then without warning, Garcia asks, "Why do you keep that man around you?"

O'Shay instantly knows he means the driver who spoke out of turn.

"Look," O'Shay says softly, very softly, "he is no longer with me. Everyone needs a pit bull in this business, a dog bred to kill other dogs…"

Garcia nods his understanding and offers, "I know what you are talking about."

"The thing about a good pit bull," O'Shay continues, "is that it keeps its mouth shut, it doesn't yap. When one of them starts yapping, sometimes you need to put them in a cage…"

Garcia nods his complete understanding.

"But," O'Shay purrs on, "if they keep yapping, you put them to sleep."

Garcia's head rises slightly, his eyes meet O'Shay's, and he says with a flat voice, "I know what you mean."

"Now that dog is in a cage," O'Shay rolls on, "but if he keeps barking, I'm gonna put him down. In my business, I cannot afford for someone to lie. Maybe in his stupidity and ignorance, he thought I wanted to impress you that I had jets."

Garcia stares dead-on into O'Shay's eyes.

"Let me assure you," O'Shay continues, "I have no fucking jets. If I had a goddamn jet, why would I need you to come here? He was out of line to tell you such horrible lies, and you have my apology. All I can say is that what he told you is a lie, that he tried to make me out as something I am not.

"I will show my cards to you: it is obvious that I need to get a good product here. The boy who did the yappin' knows enough about my business that it is my fault that he even still works for me. But what I don't appreciate is that now you know my needs and you will be able to take advantage of me because you know I have to have this product.

"If you choose not to do business with me, that is your decision. I won't blame you and maybe we can spend your remaining time here fishing. But if you choose to do business with me, I promise you that you will make money. But I apologize for the behavior of that person."

Garcia looks at O'Shay and says nothing.

The next day O'Shay goes to Garcia's hotel and he can sense instantly that it is on. Garcia has clearly made calls to Colombia. Garcia says he likes the idea, that this city is the base they need.

O'Shay makes some calls from Garcia's room on a special phone. He explains to the Colombian that the phone will work for thirty days, that the number has been lifted and calls can be made anywhere in the world and there will be no bill, certainly no bill that ever comes to whoever is using the phone. And then after thirty days, the phone company will realize what is going on and kill the number. Garcia is fascinated, and O'Shay sees a gleam in his cold eyes. So he hands the phone to Garcia and says, "Here, it is nothing, you should have it."

Garcia picks up the phone and a small smile graces his face.

O'Shay excuses himself for a moment and returns with a map, spreads it out, and says, "Okay, here is what I want to do. Small planes can get it here, and I can move it out into the country. But what you want to do is front me dope because if you front me dope then you are part of my organization. If I'm just buying it from you, then you are not. But if you are willing to give me what I need at the price I hear you can, well, then we can make big profits."

Garcia seems taken aback. He is sitting on the couch, O'Shay is in a big chair facing him, and Garcia's face seems almost to drain of color, his eyes tighten and he looks suddenly frail compared to O'Shay's bulk.

He says, "Well, my man in Colombia wants some kind of collateral, some basis for trust. He wants to know where you live."

O'Shay suddenly snaps alert, his shoulders hunch forward, his head comes up, eyes small darts on his placid face.

"Look," he says rising, "I ain't giving you a damn bit of collateral. You'll kiss my ass before I'll ever give you my address or any other fucking thing. If you want to front me stuff, that's your business. But first we go cash and you don't tell me a fucking thing about how I do my business."

Garcia says nothing.

AT TIMES WHEN HE GOES OVER the deal in his head, he can feel fingernails on a chalkboard and the screeches of these nails capture his frustration. He makes a plan and then he must endure the mauling of his plan as others paw it. But he eventually snaps himself back into a kind of ease with this fact. And he can do this because he knows his crew and he knows something that at times gets lost in the fury of moments. He trusts them. He trusts them to risk their lives for him. He trusts them to never fold, never cut and run. Never break.

Joey O'Shay has gone through countless doors with his gun drawn but he has never had to watch his back, his crew, his colleagues, were there and protected him. At such moments, he is never alone, he is barely an identity, he is part of an organism and this organism is more precious than life itself. It is the surrender in those moments to something grander than his ego, to being part of a pure thing. And at such moments there is no main man, there is no Joey O'Shay, there is simply this group . . .

and Christ I cannot even say this out loud, they are brothers in arms, six fucking guys, Skipper, a Creole out of Louisiana, a guy protecting my ass, hell, the guy supervising me and fending off interference from the brass, a hard motherfucker, believes in me and my voodoo shit, Angelo, an Italian kid who walked away from a shot at pro football to be a soldier in this shit, a guy who has seen too much death, Matteo, a tough kid from the streets and he never questions me and would follow me into hell, Sonny who always laughs and yet thrives undercover and drinks down endless hours of work and is a dead shot and Tye who can hear the whisper of real information in any wiretap and who I turn to for reality checks and Sky who seems to be my twin without the curse of my evil, driven like me, and the driver who talked too much to Garcia and had to be put down to save the deal and who never flinched when risk brushed against his face and yet I love him as I do the others as . . . my brothers in arms . . .

and here O'Shay breaks off his internal mumbling to himself, but then it comes back again, because he cannot escape life itself where you get born and you die and you seem alone and yet you know a baby never held becomes a fucking idiot and no one does a damn thing alone, and that every writer who thinks the book is a solo act dies a fool who is ignorant of the multitudes driving his pen across the page and Jesus, O'Shay can't get it straight except in his guts and in his guts little fuckups drive him crazy and yet when he goes through the door he is never alone, he really never is even Joey O'Shay, he's part of this brotherhood he cannot name, this living thing he has joined and not made his life but made his only shot at tasting life.

Ah, shit, he can't sort it out because it is like a weave and you pull a thread and keep pulling and you wind up with nothing but a pile of string. The thing is the weave itself. O'Shay paces through his work and his deal and he is alone but he knows he is never alone. Alone means being dead and he lives in a web of support that keeps him alive.

And the others, that straight and narrow world, they will never know because they never put their life on the line, never bet everything. That's the real alone part, being part of a crew that voyages into lands that cannot be reported or believed.

THE NEXT DAY O'SHAY MEETS HIM for coffee in a nook of the hotel where they can be alone.

Garcia offers, "I am told you have people who can take care of problems."

O'Shay nods.

"I got a problem with some tires," he offers and O'Shay slowly realizes he must mean blacks, "that are not paying. Can you eliminate them? They are in Boston."

O'Shay nods again.

Garcia moves on to his problem with the priest in Miami.

O'Shay says Alma has told him of this situation.

"Do you want to eliminate a priest? I don't know if I can do that."

But Garcia persists, explains how the priest had taken shipment and refused to pay, keeps returning to this one problem, and O'Shay senses two things. That Garcia is personally responsible for this load and the lack of payment. And that he is puzzled about what to do, that in some inner Catholic recess, beyond the destruction he faces if he does not make good on the load or the serious law enforcement he will draw down if he murders a priest, he is troubled by a possible fate worse than death.

The death of the padre hangs there, O'Shay says nothing more. His silence says, We shall see.

O'Shay offers, "I do not hurry and no one should ever hurry me. The reason I have existed so long is that I do not hurry."

The Colombian tells him he will be calling, that in New York he will make arrangements. He wants O'Shay to come to New York and O'Shay says nothing. His eyes announce that he makes his own travel plans, moves on his own schedule.

But the heroin cannot sit there, it cannot be passive. Once it exists, it must move. That is why it exists.

There is no out.

Garcia leaves by train with no explanation. O'Shay thinks he must be carrying. But he does not ask. Nor does he take him to the train station. Other people do things like that.

The train rolls toward New York.

THERE ARE THREE PLANETS, each with a ring. One is painted with the markings found on a bombsight, the circles and hash marks that guide a lethal weapon to its meat. The next is painted with camouflage, a celestial body ready for war. The third is painted with the Stars and Stripes and the ring embracing this planet is red.

The three float in a gray cosmos, a cold and chilly place, and next to them floats a skull and this face of death is larger even than the planets themselves. In Cosima's universe, God is not in the heavens. She knows better than that. The priests, they all lie. And want to touch, also.

Splashed below the three planets and the skull is a map of the world with all the continents and both poles clearly sketched. This is the world she does business in and it is a dumb and unknowing world, a place where she feeds and sells and kills and no one seems to notice. A world that does not understand the real dangers as she does. And this includes Joey O'Shay. He is bright and dangerous, but in the end he has that red-white-and-blue problem. He does not grasp how evil the universe truly is.

And she, standing there with her brush, she does.

O'SHAY IS IN HIS MID-TWENTIES in uniform, the moon is full, and he knows that anyone who says a full moon does not matter is out of their fucking mind. He's in a bar of Mexicans, an all-night place, the night is hot, a sandstorm blows in, and the moon goes red. Then a hot rain falls and mixes with the dust.

The shooting starts between two Mexican gangs just down the block, and two die and one is paralyzed and all the survivors look crazed from the heat and the dust and the booze.

O'Shay feels the world has gone mad and the moon has gone mad and then he suddenly feels something stranger, something cold in the midst of the heat, the suffocating heat: he looks behind him and sees this form with eyes, burning eyes, and he thinks, Ah, it must be a huge fucking dog.

He asks his partner, "Do you see that?"

"Yeah."

And then O'Shay catches that smell of the Evil Creature. But a flashlight reveals there is nothing there.

O'SHAY MAKES IT A RULE to know and when Angel Sonrisa goes down, he forces himself to learn what happened. Sonrisa, that smiling face, that big walrus mustache, that happy-go-lucky disposition, this Sonrisa, his mentor, did not make mistakes. Did not hesitate. Did not miss anything.

Sonrisa had gone to do a deal and they cut him in half with machine guns before he even got out of the car. All Sonrisa had time to say was "Oh, hell."

There is no way to prepare for such a moment, O'Shay thinks.

Sonrisa rolls up in his car to do an undercover deal and can't unbuckle his seat belt before he eats a barrage. Not even an ice pick would have mattered, O'Shay thinks.

And it all means nothing.

Once when O'Shay was starting up, he went dove hunting with some guys in the country. He drove out very early in the morning, while the rest of them had gone the night before and slept in a barn. When he gets there, the one guy has his two kids along, a girl about sixteen and a boy about three. Their bedrolls are still sprawled across the floor of the barn. The guy is already drinking and O'Shay notes this fact because he can't abide drinking before noon.

They all stand there and talk a bit before they head out with the shotguns to get some doves. The three-year-old wanders around the barn and no one pays much attention. He finds a .38 in one of the bedrolls.

It is a fine September morning, the day will surely run one hundred degrees and O'Shay looks down and there is a three-year-old holding a pistol a foot from his chest. The kid has both hands on the gun and is trying to squeeze the trigger. His aim is dead center.

Without a word, O'Shay reaches and grabs the gun from his hands.

Everyone laughs.

O'Shay laughs, too.

Sonrisa rolls slowly down the street heading for the meeting. He's got his knife, his gun, his eyes.

He pulls into the parking lot. Lights splash across the area. Still got his seat belt on.

There is a loud noise, and Angel Sonrisa says "Oh, hell," and then he is gone.

O'Shay has seen too much to ask why. But still, he asks why.

But none of this matters because Sonrisa knew more than O'Shay and still he went down and being the expert afterward misses the point.

He goes years at a time without thinking of that morning in the barn with the three-year-old, simply erases the memory for long periods because the memory tells him nothing about being safe, teaches him nothing about not relaxing. The memory simply tells him that no matter what you do, how alert you become, you cannot be completely safe. Safer, yes, but not safe.

O'SHAY RETREATS INTO his comfort zone. The images dance on the screen, the slow-moving half-wit fiddles with the motor on a lawn mower and Joey O'Shay can smell justice on the wind rising out of hell. The story line is pleasing, the mother and her lover slaughtered by an outraged son. And then the long suffering inside the walls of the state, and now a freedom, one that looks a lot like the green and open possibilities O'Shay knew as a boy along the creek. And floating on his jon boat in the cove under the stars.

The white-trash texture of *Sling Blade* feels right also. The demon boyfriend with his booze, the woman trying to maintain, and the child, the child with innocent eyes seeing evil transgress his world.

Everything is fine in this film, a dead mother, a worthless father, suffering, and now the last gasp of innocence. And best of all, the father is played by Robert Duvall who entered films as a tortured young soul in *To Kill a Mockingbird.* Now he is older and coarser and no longer the savior of children but a menace to all that is good. O'Shay can smile at this flicker of movie reference and how it tracks the corruption that life doles out to everyone.

And of course, the retarded hulk, Karl, has his attractions for O'Shay. He is a killer, he is maimed by life, he is dangerous. But he is in his soul essentially good and when the moment makes it necessary he will protect the child from harm. O'Shay loves movies with innocence and the stench of evil choking the air around this innocence. He craves the hopes that all children have and he feels the failures of all adults to protect those hopes. And to protect the children.

He can feel the movie build to the ending it must have and this time the terrible sword of justice is a thinner, sharper weapon but a blade all the same.

The movie is part of O'Shay's medical kit on certain evenings, a thing he can reach for and feel himself partially restored.

7

E IS FLYING. The city he leaves bakes, the city he catches off his wing is new to him. O'Shay does not travel often, he makes things come to him. For two or three months, he has been on the phone with Garcia. They talk about fishing, about life. About cartons, various cartons. They never mention the word heroin, they never mention murder contracts. Cartons, matters to clean up, words like that. They never mention anything concrete except for fish and women. And now, O'Shay is flying to move his deal along.

As the plane comes into LaGuardia he is stunned by the way the city rises up, a row of spikes resting on some platform and surrounded by water. He catches a glimpse of the Statue of Liberty and is surprised at its size. He feels awe. The statue is huge, green, beckoning. The river—East River?—yeah, the river is gold and rolling. And he looks off the wing and sees beauty and then thinks there is beauty even considering what has happened here.

His city is a big city and his city has a beautiful skyline, yes, he tells himself, his ground, also, has beauty. But the towers of his city would be lost in this one. The towers here go on for fucking ever and my God, he cannot stop staring at the beauty, the spires, the golden river, the green goddess rising from the waters.

Then the plane sweeps down over a bay and, Jesus Christ, he wonders if they're going to ditch at sea or something, but he notices the other passengers are looking out the window and no one is hollering, so it must be okay. Then he hears a thunk, the wheels touch down, the city skyline disappears, and he is on strange turf to make his deal. He relaxes. He lets care go. The deal is his, his if he wants it.

The airport is smaller than he expected, kind of small-town almost. And the people, he's braced himself for New Yorkers, but hell, they don't seem any ruder than the assholes he's used to dealing with. A guy in a suit walks up to him, looks kind of Russian, says, Want to buy a ride? and so O'Shay is in the back of a town car rolling down to the Embassy Suites at Battery Park.

He leans back, the density of the place washes over him, the tenements with flags draped off balconies, the tunnel, then Manhattan, the compression of life into concrete, brick, and stone. He lowers his window, lets the smell of water and garbage and exhaust fill his nostrils, his eyes tighten as they hit the city streets and the shops flash by, the sidewalks bloom with people and he cannot help but look at the faces, even though he is looking for nothing. He always goes to the face and especially the eyes. He has taught himself to watch eyes. The day is cool, crisp, the light brilliant, and O'Shay flows into the parade passing his eyes, hypnotized by the colors and forms.

The driver asks, "You been here before? Where you from?"

O'Shay keeps staring out. He can feel the driver's eyes and sense he has made him for someone in the life.

At the hotel the reservations clerk asks him where he is from and O'Shay comes up short and thinks, He must hear some kind of accent in me. Even the fucking doorman asked where he was from. The clerk says, "Do you want to face the river or the other way?"

He rides up the elevator, walks into his room, sets down his suitcase, retrieves his gun. His men are stashed in other rooms.

He does not want them here at this moment. The curtains are pulled shut. He walks over and snaps them open.

He thinks, I have no fuckin' idea what I'm going to see. And he looks down into the hole, the place Cosima keeps making her fucking paintings about, the footprint of the vanished towers, and it hits him, he starts crying, he can't explain it to himself, he feels sick. He suddenly realizes that what is gone was twice as big as the tallest tower in his city, or damn near twice, this is horrible.

This is horrible because ... and then the words stop. He doesn't trust words anyway, but now, for a moment, O'Shay knows the words can't reach out and touch that hole.

He sits and stares.

He needs to touch something solid. The city is more than he expected, the river and statue also. And now the hole. O'Shay must find his focus again, he can't be at the mercy of feelings.

He picks up the phone and tells Alma where to meet him, he checks his money once again, seventy grand for a sample kilo of heroin, a kind of test for quality before he commits to shipments. He spends the evening reviewing details. And pulls the curtains shut.

They meet the next day at a midtown hotel. When O'Shay arrives, he finds the place teeming with creatures from television, some kind of grand occasion has taken over the building where he is scheduled to sit down with the Colombians. There are lines everywhere and what look to be thousands of fucking cops, he notices. Garcia is in the hotel and O'Shay calls his room and says, Let's meet in the bar.

Garcia comes down with a woman. She is old, and has a blond wig and a hard face.

O'Shay's men are with him and he tells Garcia the money is in a bag. The old woman smirks.

Slowly, O'Shay understands that Garcia has not brought the heroin.

He says, "I've come all this way to buy that kilo and you don't have it here? I don't know what in the goddamn fucking hell is going on here. I need to get this and get the fuck out of here."

The old woman stares hard at him. The bar is clatter, outside the streets packed with cops and celebrities and media. O'Shay does not like the feel of the room.

He turns to the old woman and says, "I want to explain something to you. I don't know who in the hell you are. I've come here because Garcia said he would have this thing for me."

"Where is it?" the old woman suddenly asks.

O'Shay's colleague, Matteo, puts the small bag on the table.

Garcia and the old woman stare at the bag with greed seeping from their eyes. O'Shay can feel their hunger. They want it and he knows by their desire that he will control them.

"There it is. It don't mean shit to me. I come over to this hotel and I can't even check into a room and take a shit. There's cops everywhere because of some goddamn convention."

The old woman starts laughing.

O'Shay looks at her. "I don't like surprises."

She asks, "What kind of people are you?"

"We're redbones, mixed-race people from the rivers and swamps. We're not attached to anyone."

The old woman says, "I like that."

She points to Alma and says, "Send her with me and I'll take her to it."

After they leave, O'Shay is ready to end the meeting. There is something off, he can smell it and he wants out.

But Garcia motions him to stay and beckons to another woman in the bar.

She wears a tailored gray pants suit, has a lush body sculpted by surgery and tens of thousands of dollars' worth of jewelry glowing on her olive skin. A scent like vanilla floats off her.

She is introduced as Irma. O'Shay can immediately sense

that Garcia works for her. She has an air of confidence, of command, of having men bend to her beauty, and O'Shay appreciates this fact, and admires her craft and bearing.

Matteo starts to flirt with her and O'Shay says, "See, he's falling in love with you," and she laughs. He can feel her ego, firm and large like her breasts, and he is comforted to sense her weakness.

He decides to stop wasting time.

"I'm here to do business. I can handle ten to twenty kilos at a time. Delivery every two weeks. I have cocaine, I move it from the border along with marijuana, hydroponically grown stuff. I move things in small airplanes."

She says, "We can do business, yes, we can do business."

And then she smiles and O'Shay sees her teeth white and gleaming and takes in that scent, he can't quite nail the scent, a faint taste of vanilla.

"I am connected in Colombia," she continues. "I want to stay in the United States. We need to meet people like you, we need transportation. I want to come to your city."

He orders a beer, she has an Amaretto sour.

They go over possibilities. Irma gives no background on herself. O'Shay realizes the trip has been a test, that he is being studied, that the old hag is their bloodhound brought in to analyze his scent. He can see the old woman's life, sucking cocks in alleys and then setting them up to be robbed and killed, working her way slowly into the light and warmth of the room, her body wiry, jaw square, yes, that's it, he thinks, eyes a hazel green, a whore, a snake. He savors the instant when they locked eyes, he could feel her meanness and wanted her to feel his.

He knows that when Alma returns, she will have no heroin. He knows she is on a fool's errand and that eyes will watch her, and questions will seek weaknesses. And suddenly he feels the edge of boredom. He is here in this bar with Colombians, and yet he is watching it happen, waiting out the inevitable moves and words. He is alert but not completely engaged. The hotel is noisy, the police roam, the Colombians sip their drinks and make conversation, Matteo continues to flirt with Irma.

Alma returns, O'Shay stands. There is nothing left but business, and if there is no business, he is leaving.

That night he dines in Little Italy, he rolls the day through his mind, he takes in the city's scents and sounds, allows himself the moments of a man visiting to no purpose.

The next day, Garcia calls and says he cannot get the kilo, that nothing available is good enough, and for O'Shay it must be the best of the best or nothing.

O'Shay says, "Well, we're going back home."

A day later, they deliver the heroin to Alma.

It has been a series of tests. When the old hag took Alma to the stash house, she peppered her with questions about O'Shay and his business. Alma said, "He came here for the product and if he does not get it, you are probably going to lose him." The Colombians at the stash house asked no questions and were warm and friendly, but the house was bare wooden floors and emptiness. And O'Shay, when he learns this, realizes that this was another test, to see if he would rip off their stash house, and he takes comfort in this caution because it shows that they are stupid, and that they have no real sense of him. He is someone they cannot quite pencil into their world.

He feels control. The heroin connection will come into being or it will not. It does not matter. The venture will make money or it will not. Garcia will arrange for major shipments or do nothing. Never hurry, never want, never hesitate to walk away from a deal.

He is back in his own city, he has his own work. New York seems a series of tiles suspended in space, the gold of a river, the garlic in a café, the waves of sound, people walking, a woman with a perfect body drinking while talking loads and prices, a blue gym bag holding cash resting on the dark wood table in the hotel bar, the face of an alley in an old woman's eyes as she studies him like an insect.

He tries to clear his mind of the trip. Garcia calls, he is going to Colombia to arrange things. Fine. Irma insists she will visit O'Shay in his city. Fine.

O'Shay has spent years getting to this moment, slowly, patiently building his world until he can reach out and touch the cords of commerce on other continents. He has mastered guns but is not a great shot. He has taught himself Spanish but is not a great speaker. He has taught himself all aspects of his business and now he strokes the webs of those who are considered the best, the heroin people, and it does not seem enough.

THEY STUMBLE, GUARDS SHOUT, the rocks are big, they stumble on. They walk through puddles, they are wet and cold, the guards slam the butts of their rifles to keep the men moving. The wind is ice.

Viktor Frankl endures.

The person next to him whispers, "If our wives could see us now! I do hope they are better off in their camps and don't know what is happening to us."

O'Shay has trouble as he looks down at the page. He can smell the hair of a child, that scent that is not really a scent. He sees that boy suddenly, the anger comes back—yes, if the wives could see us now.

Frankl slips on the ice, stumbles forward, and his wife dances before his eyes. Sometimes in the gray of morning he can see the stars and he thinks, Somewhere out there my wife is under the same canopy of heaven and she is smiling and her smile outshines the sun rising.

He sees the truth and thinks that with all his learning this is the first time the truth has been revealed to him, "that love is the ultimate and the highest goal to which man can aspire. Then I grasped the meaning of the greatest secret that human poetry and human thought and belief have to impart: The salvation of man is through love and in love."

O'Shay feels the boy slipping away again.

Love.

RABBIT IS A CARTOON VERSION of a biker O'Shay knew in his own speed freak and biker years, maybe five nine and a scrawny 135 pounds, a wiry piece of shit wallpapered with tattoos. His old lady runs more along the lines of concrete block, maybe five three and a muscular 140. And she carries the only full set of balls. She likes armed robbery.

So O'Shay is over at their apartment, and Rabbit's calling his old lady a whore and she's cussing him back. He's drunk, sitting on the couch in boxer shorts, a tank top on his white hide and flip-flops on his feet. She's wearing shorts and cowboy boots when she decides she's had enough.

Her first kick hits him in the shin and when he topples off the couch onto the floor, she gives him another three or four in the face and guts. O'Shay grabs her and feels her strength and she breaks free and starts smashing Rabbit's face with her fists. O'Shay grabs her again and throws her outside and locks the door.

The next morning, he comes by to see how Rabbit is faring after his beating and some guy is there, some friend of Rabbit's, and this guy is acting big, kind of throwing his weight around.

He asks O'Shay just who the fuck he is and O'Shay does not like the way he asks the question.

So O'Shay bitch-slaps the guy, kicks him hard right on his spinal cord, then pushes him out onto a balcony and watches as the asshole goes over the rail and falls a story to the concrete below.

O'Shay goes downstairs, finds the guy alive but moaning. He rolls him over with the toe of his boot. Rabbit has come along and when he sees how bad the guy is, he gives him a couple more kicks for the hell of it.

O'Shay leans over and whispers one thought into the guy's ear: "I'll kill you if you talk to the cops."

Then he leaves.

Nothing ever happens.

Including regret.

THERE WILL BE A ROOM and he will go to it. The chamber will be of stone, soundproof and stout. A shallow pool of water smothers the floor in the darkness and a soft raft awaits. Joey O'Shay will get on the raft, eyes to the ceiling, and drift. The darkness will be infused with stars reflected on the ceiling by a fiber optic dot. He can go here to heal and master the emotions raging within him. The air will sag with scents—willow or rain or honeysuckle. A lever will let him select whatever intoxicating fragrance he desires. Even as O'Shay thinks about it, the force of a spring rain slaps him in the face, the lazy love of honeysuckle on an August night caresses his senses.

The stars will be of midnight on a moonless night, explosions of light from the deep space of the place called the universe. They will rescue him from the emptiness of this planet called Earth. He has never felt alone when he sees the stars.

For years, he has planned this chamber, a sanctuary no known drug can replicate. He is convinced that in such a place, he will heal. Not be made anew, but heal, gain strength, recruit what he needs in order to continue.

Nothing can erase what he has done or what he does or will

do. Nor does he seek such an obliteration of his path. That is another matter. He wants to drag with him all his baggage, huge steamer trunks of pain. He wants to remember everything. He wants all the faces to stay with him.

He does not want to be freed of guilt. He is not even sure he feels guilt. But there is this weight, something far heavier than the pain. He wants the strength to carry this weight.

So in his mind, he builds this stone chamber, the stars emerge, willow rolls across his face, saplings rooted on the banks of some stream that carries him along.

FOR COSIMA, IT IS ALWAYS about money. Officially, at least, it is always about money.

O'Shay is up to something without her and this will make her no money. After she made the initial calls and connections, he seemed to drop her from his life. She thinks maybe it has to do with the dark Dominican woman, Alma, yes, with Monkey Woman. Cosima set O'Shay up with Alvarez and then Cosima was suddenly out of the picture and Monkey Woman has taken over what should be Cosima's show. She knows Alma has no use for her either and calls her Miss Piggy. But that is not the issue. Hell, she has known Alma and her ways since they were both in the same prison. It is money. And it is O'Shay. She is bound to O'Shay in a way he does not seem to understand.

She starts calling his phone constantly and he does not answer, or he answers and calls her Puta, Whore. She ignores this. Cosima prides herself on her sixth sense, this thing gringos lack but which she possesses. She can read eyes, search souls, catch rumors on the faintest breeze. Hear the messages of midnight.

She speaks to O'Shay with a smirk on her face that constantly whispers that he has no hope of pulling off this deal without her. That he needs her whether he realizes it or not. That she should be part of his deal and there is no way he can stop her from being a part. That the deal exists only because of her and her connections.

The phone rings, O'Shay answers. It is Cosima breathless with a new Mexican heroin connection. He brushes her off, he is tired of dealing with low-grade product, he is intoxicated with the idea of pure Colombian. She persists, call after call, probing, pleading, convinced she will wear him down.

She fails, and she cannot accept this failure.

So she goes for her ace, the one asset she has always kept from him, has hoarded like a kind of private treasure. She reaches back to her prison days.

Cosima is doing three years federal, and it is her second fall. Alma is already in that prison, and she watches as Cosima enters, a fresh thing, and dyke eyes wash over her. Her body is curves, her breasts large, hands small, face like a doll's. She is ripe.

Cosima's cell mate takes her under her protection. The woman is also in her thirties but nothing like Cosima. She is a woman who does not need men but focuses on her son and daughter. She does her time by working out and reading. Her face is soft, her dark skin and black hair echoing the mixed blood of the islands. She speaks softly and without vulgarity, but she is feared in the prison because of her connections, which reach deep into Colombia heroin and the Italian Mafia. She moves like a schoolgirl, her eyes dance, her hips loose and yet innocent, that toss of her dark hair, the smile clean and fresh.

Her name is Gloria, and for three years she protects Cosima, makes prison for her *muy tranquilo*, a quiet island in her life of noise. Gloria tells the blacks to leave Cosima alone, gets her out of working, shelters her when her constant thefts arouse anger and the possibility of reprisals. She even fixes things when Cosima gets in a fight with a guard. And Cosima returns the favor. She gets a hot plate and makes special desserts for them to share, like two little girls having treats.

When they both get out, they stay in touch like old friends. She is the one person Cosima keeps in a separate category, not as a connection or a business associate, not as someone she rips off or has murdered. Gloria is almost the only thing in Cosima's life that is not exploited, used, and destroyed by her appetites.

Cosima for years has kept Gloria from O'Shay's reach, kept her pure and inviolate, an act she cannot explain to herself or find precedent for in her life. She has watched men be executed and not flinched. She has sent people to their deaths without a thought.

Even Bobbie, who needs always to prove that the world is a scam, balks at what she senses in Cosima, a blackness like evil, something like what she senses lives inside O'Shay. She feels foolish even using that word, as if she has stumbled into a dead language. It is a word best left to preachers.

Evil came crawling from the dank mists of northern Europe, *uvel* in Middle English, *yfel* in Old English, then *ubil, evel, upil,* and finally *evil* as the word journeyed across various tongues. It began as meaning "exceeding due measure" or "overstepping proper limits." Sometime in the nineteenth century, it largely left the spoken tongue and retreated into literature, where it was replaced on the street by the word *bad.* But bad simply is not bad enough for some things, not bad enough for willful destruction of others, not bad enough for joyful cruelty, not bad enough for Joey O'Shay at times, or for Cosima.

Evil seems necessary even if unspoken, seems the only way to explain ruining people for no other reason than appetite, ruining others not in order to survive, not because of some greater good, not for any reason save because such ruin is possible. Evil becomes a necessary tool for some as they struggle to make the world seem intelligible to themselves. No matter how they live, O'Shay and Cosima need for things to make sense, need for the paintings to exist, the paintings of a duck landing on a safe pond at dusk, paintings of two towers falling from the sky.

Cosima will be purring on the phone, coaxing a connection into a delivery, O'Shay will be plotting the deal of a lifetime but neither of them can function if it is only about drugs and money. There has to be another dimension, one unspoken but felt, a place or terrain that takes them past the workaday schedule of eating and phoning and meeting and counting out money. Cosima thinks about Gloria, sees O'Shay moving beyond her reach. In prison Cosima, Gloria, and Alma were together. Now Alma has led O'Shay to Garcia, and so Cosima is about to make that call to Gloria, and Alma will not know of this move just as Cosima will not tell Gloria what Alma has arranged for O'Shay.

Cosima tells O'Shay that Gloria is interested and wishes to come to the city. He hesitates, wondering why he needs yet another heroin connection, golden as Cosima's connections are. He can taste Cosima's surrender, savor the fact that she is doing exactly what she has always refused to do, bringing her most cherished connection within his orbit. That she is joining him in this place where he lives, a place he can only describe with that word, evil.

So he thinks, *What the fuck. I'll annihilate this bitch while I'm doing the others. I'll order up some more heroin.*

O'Shay has Gloria flown in first class from New York.

He has Cosima pick her up so as not to lose face. Then he almost ignores her. The second day, he drops by and visits her in the evening, and they talk about the trip, the weather. O'Shay tires of this and says, "Look, what I need is a good heroin connection."

Gloria nods. She can see his eyes on her.

She says crisply, "Everything with me is business. All business. I don't need a man in my life. This is business for my children."

He takes her in, her smooth complexion, her large breasts, the curves of her body. She smells like some tropical fruit and she moves slowly and with grace. Her head cocks, she tosses her black hair, then comes the smile. A slight overbite that makes her sensual. Her hands are perfectly manicured.

O'Shay hands her and Cosima some money and says, Go out and have a good time.

She leaves the next day. But first she gives O'Shay a phone number.

He calls her a day or so later to make sure her flight home to New York was satisfactory.

She says, "I want to tell you something. I like you, and you were very polite and kind to me. I'm going to put my daughter on the phone so that she can explain something in English..." and then suddenly, the college-age daughter is speaking into O'Shay's ear that Cosima is scheming to cheat him by artificially raising the price and that is not the way her mother does business.

O'Shay takes this in, the fact that Cosima is secretly gouging a little extra out of the deal for herself, and replies, "Tell your mother that I know Cosima is greedy and sneaky and lazy, and that I have to keep an eye on her. Tell your momma I won't do anything to Cosima," and he can hear Gloria laughing in the background as her daughter translates his words.

O'Shay continues calling her, calling her contrary to his habit of having brief, clipped conversations with connections. Sometimes he gets her son or daughter on the line and he falls into the habit of talking with them, though never about the business. They are the children of a major Colombian player but have been largely shielded from the business by their mother. Slowly he gets to know them, even though they are only voices, gets a feeling for them.

Gloria falls sick, a bad cough, and O'Shay is relaying medical advice, insisting that the daughter take her to the doctor. He fears she has pneumonia, and he realizes he is slipping into something that is not a normal business relationship.

She once tells him, "I know I'm fucking up because I care about you."

Soon he has all of her phone numbers. She juggles six to eight phone numbers, changing them constantly, but she always alerts him when she makes such a change and so he is always in her world and flooding her ears with conversation.

He can feel this slippage in him but he denies it. Just as he thinks of her body but denies this fact. He knows from Cosima that Gloria's body is extraordinary, Cosima tells him how they would shower together in prison and how soft and yet firm she is. He senses a softness inside, too, as if she is protecting some part of herself that has been damaged, harboring some dark experience that still wounds her. She has the aura of a woman who will never trust a man again, who is determined to be independent, and yet she keeps talking to O'Shay, almost girlishly, about how she likes a quiet life, loves old movies, loves her children, loves simply staying home, curled up in bed with old love stories streaming across the screen. He thinks of her as regular folks but he can do the numbers in his head and know that she is clearing at least a million a year. Still, she is the homebody to him, the woman who has sheltered her soul from the business.

Both of them disappear into this fantasy of quiet domestic life and soft words and perhaps cups of coffee in the morning, of a life that is not business, a life that is not harsh, the kind of place heroin promises but never delivers. They go there against their will and yet willingly.

In the sky over the city, satellites scan the air, on the ground giant dishes suck down information. In the capital, analysts pore over conversations, numbers, seek patterns. And they begin to discover calls from a phone in New York, a phone that reaches into Colombia, the islands, Mexico, a phone that touches places along the U.S. border, hits known traffickers in major cities, a phone that radiates out into the world of heroin.

And this phone is in a nice house in Queens, New York, a house in a good neighborhood where homes are expensive, a house that they discover has no mortgage on it, and this phone constantly calls one number in a distant city. And so they check that number, too. And find this link between a known heroin dealer named Gloria and someone at a phone in the other city named Joey O'Shay and this person does not change his number. So Gloria comes to their attention.

The analysts in the agencies are about patterns, and others can pursue these patterns and run them into the ground and eventually destroy those who are caught within the patterns. They realize Gloria is one of the best-connected heroin brokers in the world and now they have all the threads of her world in their hands. The agencies buzz, and a plan emerges for wiretaps in Colombia and Mexico, for a leisurely exploration of Gloria's world. And this number she keeps calling that belongs to Joey O'Shay.

He is always alert, he will break off a casual conversation if he hears the slightest noise, even a voice on the street that penetrates a room he is in, he will stop dead until he determines what that sound is. He knows better than to use any phone near the border where the agencies all listen in to everything. He is leery of any form of communication. But he keeps calling Gloria and talking about nothing and talking and talking and talking.

O'Shay probes into soft areas of Gloria's life, explores common passions, this thing they both have for old movies, especially one O'Shay watches endlessly, a film from the forties called *Laura*, in which a detective investigates the murder of a woman only to find that she is still alive and that he is in love with whoever she is and … they keep talking, the dishes keep tracking calls, the deal inches forward and they both think about how they like a quiet life, one with old movies playing softly in the silence of a bedroom.

It is all about control. Gloria hits her numbers, voices come on line, from verandas in the islands, from patios on the lip of the Andes, from Miami or Tijuana or Los Angeles, and her phone numbers keep changing, her language is a code, the television silently runs a movie as she shuffles phones and hits her numbers and keeps control of her world. And Bobbie is at her keyboard, fingers flying, numbers, rooms, rates shifting columns, the phone cradled on her shoulder, the world at her beck and call, everything moving faster and faster and with absolute certainty. O'Shay reviewing again and again the details of his deals, making his calls, never being explicit, lifting those weights, feeling his muscles flood with blood, tasting the sense of mastery.

O'Shay does not realize that satellites are beaming down, dishes are sucking up messages, analysts are pouring over a river of babble, seeking that key word, that phrase, that pattern, the skeleton key that will suddenly give them control of a world they hardly venture into, a place they know only as a rumor as they sit in their windowless cubicles pawing through the hungers they have never personally felt.

And no one is in control, not Cosima seeking her way into O'Shay's deal, not O'Shay shuffling various heroin connections, not the electronic ears listening, not Bobbie with her fast mind and fingers, no one, absolutely no one is in control. They simply limit their reality so that it appears controllable, safe, defined, concrete, make it into something that can fit into a file, or a little black book of numbers.

THE NIGHT FEELS SOFT, O'Shay shifts and the rumble of the engine rolls through him as the tires grab the city streets. It is good now, this is a time much earlier in the game, the air still has scent, the women swing their hips and dreams sing out from them. All the lips are red and warm.

The deals are everywhere, the heroin, marijuana, cocaine, all the kinds of pills people crave so that they can tolerate the city he now owns at night. And he has learned that these people are everywhere, all classes, the best neighborhoods, the finest schools, the towers where the money is sorted, the clubs where he is denied entrance, in all these places the people need certain chemicals and sensations so they can bear the city that is rolling under his wheels as the stars try to peek through the scud of clouds floating overhead after the fresh rain that still glistens on the asphalt.

The car is gold, pure gold, and under the streetlights it is a sun coursing through the black lanes. He is without an ounce of fat, the body hard, the eyes focused, his mind never stops working, nor do the eyes for that matter, music purring from the radio, the city now safe, a possession, something known and mastered, something to be caressed or trashed, taken and done with as his mood suits him. The women are everywhere, in all the districts, all the colors, they are to be visited, pit stops in his life.

No one asks him how he does what he does and he offers no explanation.

He is an independent operator, this is not supposed to be, but there it is and all the fictions of his relationship do not really hide this fact. He works with people but he works for no one. He makes money but he does not care about money. He has women but he is not about women.

The car is more of the same. It is golden moving through the hungry streets, seats low, engine at the ready, but it does not matter to him. Its engine breathes power, and that is what they see. But he is really commanding something else.

And that, they never see coming.

That night, that time, it was very good then.

Before he got better and knew too many things.

THE TILE OF THE BATHROOM that surrounds Popsicle is plain, not fancy tile, but plain. It is clean, someone cares about simple hygiene. The bathroom itself is not fancy, the sink, the tub, the tiles on the wall. No one has created a fantasyland in this bathroom.

Popsicle can breathe now. The plastic bag is gone.

She is looking up, and she looks very, very small.

Her hands are tied in front of her.

Her skin, that rich brown skin, is everywhere. Her blouse is gone, the shorts also, the bra, the panties. She is nothing but a sixteen-year-old girl standing naked in a bathroom with a very bright light shining down on her.

She is back to pleading now.

She is asking someone not to do this to her.

She is asking someone why they are doing this to her.

Her small breasts have dark brown nipples floating on them.

The tape abruptly ends.

O'SHAY IS TALKING WITH AN EX-CON, and hearing out a business proposition. Trucks, yes, and O'Shay will provide the security edge and there will be a lot of money. O'Shay wants out of his life, out of the drug deals. He thinks, Yes, this will be good.

Behind O'Shay's office is a creek bottom with trumpet vines crawling everywhere. The place feels wrong. It is four a.m., rain drips off the vines, O'Shay and Matteo catch that smell, the Evil Creature. O'Shay puts his hand on his gun at his waist, he stares into the vines and sees eyes burning. It looks like a big dog, a dog as tall as a man. He steps toward it.

There is suddenly a noise, half howl, half moan, and the thing vanishes.

Matteo shudders and says, "What was that thing?"

O'Shay says, "It shows up around here sometimes."

O'Shay goes home.

He knows now he will not take the deal, he will not try to make money off this trucking venture.

His nostrils are still flooded with that smell.

HE INSISTS TO BOBBIE that Gloria does not know English and Bobbie is stunned. O'Shay had her take Gloria to a male strip club and Bobbie picked her up in a limo and watched her actions, watched her listening to Bobbie's instructions to the driver, and Bobbie could tell, knows for certain, that Gloria followed what she said. In the club, at a table right up front because O'Shay wants everything done right, a waiter asked Gloria something in English and she answered immediately. They spent hours there drinking and ogling and she watched that infectious smile bloom on Gloria's face, saw her expression pick up every line in the act. She knows English, goddamn it. Bobbie wonders if O'Shay is losing it. She worries that he cares about Gloria. To Bobbie, caring is a fatal condition.

O'SHAY CALLS GLORIA. The daughter answers and they talk and then O'Shay asks her to put her mother on.

"Gloria, I'm going to Europe for a while on a vacation. I'm going to be out of touch for a while because I have some things there I have to do, things concerning my business."

"Oh, yes," she says. "I understand."

"Oh, baby, how are you doing? I just thought I'd check on you and see if your cold is getting better."

"Oh," she says, "I don't want you to get what I've got, baby."

"I just need to know what might be happening with those boxes you are getting me with those gifts."

"Oh, if I get those boxes they will only be part of what I promised you."

"That's all right, baby. I'm glad you are feeling better."

"Oh, it's a good thing I am not there, I would get you sick."

O'Shay can feel the warmth in her words and a flirtatious undertone. He knows that this woman, who has set up all these barricades so that she will not take a fall, so that she will not get hurt, has now failed to protect herself. He knows he has her.

He hangs up. He has not turned on the cheap black cassette player that simulates music from a club or cabaret. Sometimes

O'Shay likes to tell Gloria he is in a friend's bar. But not this time, no, for this call he is at home in his mansion and calling his girl.

He is sitting at a gray Steelcase desk. There is a sign by the phone he uses, hand lettered by O'Shay: SHOULD THIS PHONE RING, DO NOT ANSWER. Insurance against some dumb motherfucker picking it up. The call has taken maybe five minutes, O'Shay knows better than to stretch out any business call. Who knows who may be listening in? Nobody he knows in the business has a long conversation unless, say, they're having a fight with their wife—and that's not even personal.

He looks at the acoustic tiles that line the windowless room, barely a closet at ten feet by five feet. The chair is blue, there is one small blue shelf but the room itself is sterile. He checks to make sure the recording of the call is clear. He puts it in a mailer to be shipped overnight to New York.

He calls the agent there and says, "Your dirty call is on the way."

O'Shay knows that anyone in the agency who has any feeling for Spanish will sense the heat in the call, the emotion throbbing beneath the byplay. The federal people in his own office are already having questions about his relationship with Gloria.

He's told them, "Look, in order to get this woman's numbers as she changes them constantly, I've had to get personal."

He thinks about this and realizes that he simply doesn't give a fuck what they think. Now they can file and set up a wiretap. O'Shay has driven the first real nail into Gloria's coffin.

He wonders why he did it. He did not have to make the call, even when the agents from New York discovered his operation through sources in Mexico. The feds were baffled at all these major heroin connections leading to one obscure phone number in his city. They were stunned to find O'Shay behind the phone and even more stunned to discover whom he was dealing with. So they asked for a dirty call and he said, Sure, you can have it in an hour. But he did not have to do it.

He's part of a task force, a city cop attached to a bunch of federal agents. But that doesn't give them any real leverage on either him or his case. He's a key deal maker in his office and he knows that is why he's been allowed to live undercover for almost twenty years and why he is on such a loose leash. No, he didn't do this dirty call for them.

He savors that moment, the female agent in New York stunned when he tells her, "I can do it anytime. In fact, I'll do it right now and mail you the tape."

For O'Shay it is a normal decision, quick, strategic, and an easy decision because Gloria is a sidebar to his real deal, a simple buy/bust event. Dealing with her is like sweeping the porch, a kind of humdrum task, he tells himself. He'd tossed out just enough baby this and baby that when he slipped in mention of the boxes and when they would be ready, well, Gloria did not seem to notice the shift in his talk at all. And besides, he kept telling her how he wanted to see her. And she said softly back, Oh, I want to see you, too.

He steps out of the room where calls are taped and filed and stares at the office, a warren of desks and credenzas and agents from DEA. He walks past the empty desks, abandoned early as they usually are, rounds the corner and hits the area of eight desks and Steelcase walls where there is never anyone except him. And on one of the desks his voodoo altar burns, and beneath the altar is the black wooden coffin he has made, the padlocked box where the people he annihilates disappear.

He looks at the skulls, the faces of the damned floating in glass bottles. He stares at the seashells on the altar, his new addition of dried tree frogs, his herd of skulls, snakes, masks, one red with fangs, monkeys everywhere, the grinning skeleton face of a Mexican society woman who is known in folklore as Katrina, the image of lethal vanity, a painting by Cosima of eight monkeys in a row all drinking beer and in the foreground empty cans, a face floating in a bottle, the face of a man who is doomed but does not yet know it, no, he thinks he's doing a very good deal with a man named Joey O'Shay. Below, a monkey sits on the black coffin and a skeleton crawls right in front of the padlocked door where the doomed will eventually disappear forever. A flak jacket lies on the floor. A painting of an attack helicopter with large type walking across it which says WE MADE IT BEAUTIFUL BECAUSE IT IS THE LAST THING SOME PEOPLE WILL EVER SEE.

A poster from the movie *Scarface* stares down on this scene, the one where Al Pacino in a red Hawaiian shirt faces down a pile of cocaine. He remembers when he was dealing with that Russian for Afghanistan heroin and they got kind of tight, and the Russian insisted he have the poster as a gift. It is taped to a file cabinet crammed with dozens of videos of this deal and hundreds of audiotapes from wiretaps.

Maybe Gloria is watching an old movie now, one she has seen over and over before, the way O'Shay watches *Laura,* the one where the detective falls in love with the dead woman. And then a strange moment comes to the detective and he is forced to rethink his case.

O'Shay has come to his moment. Everything is in play but he feels like an outsider to the deal he has created.

He knows what will happen to Gloria now. It will take the Department of Justice three or four weeks to set up a wiretap and then, while Gloria thinks he is in Europe, she will make call after call after call to Colombia arranging the shipments for him and with each call she will seal her doom.

He knows Garcia is also busy setting up his heroin connections for O'Shay and that for a spell he made hundreds of phone calls on the phone O'Shay gave him, the one for which he was never to be billed, the free phone that was wired to reveal every single human being he spoke with. And he had been so enamored of this free phone that when he traveled to Colombia he gave it to Irma who remained in New York and all her connections and numbers came tumbling into O'Shay's reach also.

O'Shay's loaned phone to Garcia has resulted in the names and numbers of hundreds of major heroin people around the globe, many of them in nations where the police are less delicate about wiretapping and other means of gathering information. O'Shay, sitting in his city in his soundproof room, has made one of the deepest penetrations the government has ever achieved of a Colombian heroin cartel. And O'Shay gets feedback from Colombia, where the law reaps the harvest of information because, as he puts it, "Over there you don't really have to give a good goddamn how you learn things. You can just snatch people."

Then Irma calls O'Shay and says she wants Alma to come to New York and talk with her. So O'Shay sends Alma and tells her to say, "'Look, Joe's making so much fuckin' money now I don't know if I can even talk him into going through with this deal.' Tell her, 'Joe is now dealing with some Chinese people out of Canada. He's going to start getting China white.'"

O'Shay is counting on one factor: Irma's ego. She will never let the deal slip away.

Irma calls O'Shay and says she is going to come visit by herself. He tells her he will fly her down. Alma is waiting for her in the city.

O'Shay calls Bobbie and says, "Help me. I've got a bitch comin' in, I want her to stay at your hotel. We'll have the techs set it up for sound and video. Now, I don't want to look like I'm too aggressive with a Colombian woman and this bitch is beautiful. I want to make it look professional."

Bobbie says, "I have exactly what you need, I have a driver, an old gray-headed black guy with a town car. He'll treat her right."

So O'Shay gives her $500 cash for expenses.

It is all coming together. He is winning but it does not feel like winning and O'Shay is disappointed by this empty sensation. He feels closer to his prey than to cops. He feels closer to, well, someone like Gloria. She is brighter, he thinks, more decent. She works harder because she is self-employed and must work if any money is to come her way. She is sensitive, alert, alive. A good mother. There is something kindly in her, something almost genteel, a certain graciousness, a lack of vulgarity.

And now suddenly, in this last phone call, a quiet and persistent flirtatiousness that has never been there before. He is certain of its presence, could feel himself responding, saying baby this and baby that and he wonders, just for a second or two, if he was showing off, letting that agent in New York taste how close he had gotten, how deep he'd gone into this world she'd only know from spy satellites and wiretaps.

How easy it was for him to deliver on that casual promise, Sure, you'll have it in an hour. When he'd called her up in her office, he could all but hear the hum of the goddamn fluorescent lights as he played her the tape, yes, he wonders now just how much of his ego was involved in driving this nail, this dirty call, into the life of Gloria, oh, Gloria, baby. And he remembers the times he picked the music in the soundproof room, selecting a mild kind of salsa he knew she liked so she'd think her Joey was in a club and thinking of her and—presto!—called his girl.

And for what? It's been years since he believed in the law. He'd legalize all of this shit, even heroin, which in his eyes is the very worst, not because it destroys the body but because it enslaves the soul and creates the ghost people who wander the streets of his city with their lost eyes and now belong to some legion of the damned. The grass, the pills, even coke, to hell with them, he does not do such deals anymore. Let people have them and waste their lives, it's better than this thing he is involved in, this war on something that he can see in his neighbors' eyes, in the people standing with him on the supermarket checkout line.

It is not that suddenly his eyes are opening to what he is and what he does. No, he thinks, what is happening or beginning to happen is this: I can't find that line between them and me. I've always known where that line is and now I can't seem to find it. But he can look over his shoulder and see another line: the line between him and the people he works with, the people he works for.

This fucking office is studded with people making top dollar and working short hours. They get medical and pensions, every goddamn thing. And when he is here at night, nobody else is working except him and his crew. And when he is here on the weekend, nobody else is around but his own people. But Jesus, the people he hunts work his hours, the people he destroys work weekends. The people he annihilates work just like he does, are on their own lookout just as he is, have to be smart, have no pension. The people he hunts are fucking honest in their own filthy way, they keep their deals or they get goddamn dead.

The candles are going on the voodoo altar. He looks at the faces dangling in the bottles and notices that Gloria's face is not there, that he has never put her face there. She is the biggest single player he has ever done and yet she is absent from his altar. He turns away.

I don't know. It's insane. I've never thought about it. Maybe, because I could do it. I don't know. That's what I was made to do. I was put here to go after the most evil, sorry people. That's why I was put here. If they've been bad, I'm the priest. I put them in cages. But after I put them in cages, I realize they're not as bad as people think they are...

HE'S GOOD AT WHAT HE DOES, yes, he can't ignore that fact. This may be the only thing he's ever been really good at in his life. He keeps this thing clean and safe and when he reaches for it, the feeling is always there, the sense of knowing not thinking, the rush of acting not wondering, touching a piano for the first time and suddenly knowing, absolutely knowing how to play, that is the feeling the hunt gives him. Weight lifting, running, cooking, gardening, these all feel fine and they should be enough but they are not. Painting has never been enough. Police work is not really enough. But the hunt, a predator stalking other predators and largely doing it in his own mind, crafting the events according to the dictates of his own imagination, that is the only thing that has ever come close to being enough.

And now this is fading and he is not sure why. He is at the peak of his game but the game does not have the satisfactions it once gave him. And the cost is going up. He can feel more wear and tear from the game.

There are other matters, this tug of feeling he has when he talks to Gloria. He does not like to think about that. He knows it is not that he has suddenly learned something. More likely he can now admit something.

Everyone else around him is caught up in words he does not believe, words about waging war, keeping clean, enforcing a law, words about making cases, building a career, personal advancement, status, pay, words about money, money from the agencies or money from the deals. Like any business, this one generates an interior logic, a series of operations with colorful names, a list of targets who have gained enough prominence to be worth destroying, various products of the moment that became the focus of crusades. From time to time, new countries swing into view and become places to be denounced and blamed. Official explanations of the problem shift between those who sell and those who buy. Slogans come and go with the political seasons. None of this matters, not a bit. It is a language of lies that refuses to touch the flesh and the blood of the experience. The money doesn't matter either. It is never enough, it vanishes, no one in the business seems to hang on to the money. Things get bought, lawyers swallow trainloads of money. None of this touches where he lives.

The hunt is the key for him and it has to be good or he gets very little or nothing from the work. He sits in his city, he plays off the hungers and fears and emptiness of others, he makes them come to him. And then he destroys them.

He has never had to question this sense of being a special predator. It has always been there for him, his secret resource, his way of squaring all accounts. Obsession is an overused word, a thing carelessly said by those who have no idea what it feels like. They work late once in a while, they may toss in that rare weekend, they have coffee and idly think about what they are doing. And then they tell others they are obsessed with something. They have no idea.

Obsession is not like sitting at a bench making a shoe. It is not a place you visit. It is all you know, it is what you become. And everything you come in contact with feeds the obsession. There is no other time, no other reality, no other sensation. Alcohol simply dilutes it, briefly filters it. A day, a week, a month, that is nothing. Think months, a year or more, with nothing really in the mind but one thing and this thing cannot be described as an irritant or a comfort, it is everything, and all the other parts of life are faked while this one thing is life itself. This kind of focus is the ultimate addiction, the strongest drug because it gives the one thing other drugs never deliver. It gives meaning.

The drink in the hand, never a drop spilled, and yet the drink in the hand hardly exists. Conversations, hours of them, phones ringing, messages acknowledged and answered, and yet the conversations are conducted as if by remote control and hardly engage consciousness. The deals flutter in the air, the phone rings while the car flows with traffic on the freeway, a light rain misting the windshield, the voice in the air, suddenly the required shift to a particular deal or identity, all this flawlessly done,

English, Spanish, no matter, done without effort and without really being in the conversation, a kind of other life that is much less than life itself. The food sizzling in the skillet, the knife in hand chopping carefully, the slap of raw onions in the nostril, the leaves swishing in the light breeze of the night, stars over-head, moon rising, all this off to one side, not vital, there, dealt with but still on the margins, nothing compared to the taste of the obsession.

Husband, father, lover, all these things done but yet somehow remote, automatic things, things lacking the savor and intensity of the thing itself. A woman comes through the door of a café and the hand softly touches the gun while the eye takes in dress, body language, the nostrils suck down scent—on her third marriage, two kids, worried about her husband leaving, no serious hobbies, knows three dishes for company, buys books but never finishes them—all this filed and noted and yet never really engaging the mind, never for a second distracting the train of thought from the deal, the details of the deal, the essence of the hunt.

And never any words about any of this, not a single sentence to others, words might violate the purity of the thing, might con-taminate it with lies and excuses. Words might kill it and leave a person marooned on some godforsaken island where there is nothing but jobs and paychecks and pensions and little condos near golf courses where life dribbles away into nothingness.

Just as security is fatal, a disease that destroys the edge, takes the tang out of the experience, robs the senses and with a cloak of safety leads you into an illusion and once you are in this illu-sion you die. Whether you realize it or not. It is simply a matter of time once security descends.

To protect the experience, to be truly safe, the stakes must be raised, the risks enlarged, the game ratcheted up, until the senses come alive, every cell becomes wary and in this state of being, the great calm descends because everything is at risk and the only way to function is to drink deeply of the great calm. And the root of this calm lies in knowing that odds can be calculated, fallback positions staked out, every detail considered and mastered, and having done all this, knowing there is a zone beyond control and calculation, a place that cannot be fully scripted, and in that place, calm is the weapon that destroys others. Because they all have a script, they all depend on a safety net, they all have themselves covered. And when they discover they are wrong, they are paralyzed and go down.

Meaning is not something to examine, it is something to endure and be. It is the sure ground beyond questions and the plane one skates over with action. With meaning, one can notice and feel and act and yet never wonder about the actions.

O'Shay almost never has to go to court because his victims simply plead out once they learn of the evidence against them, plead out so readily that hardly any of them ever learns Joey O'Shay is the one who has done them.

O'Shay sits there, his voodoo candles sputtering in the empty office. Gloria's tape is on its way to the coast, her future is over. This time she should go down hard, much harder than the nine years she served for the last mistake.

HE DOES NOT KNOW WHETHER his wife is alive or dead. Viktor Frankl, moving in the icy wind, has no way of knowing. Messages are not delivered or accepted. The world he once knew has sunk beneath the soil, fallen from his reach, just as his great manuscript has gone into the trash.

For Frankl such knowledge has no value. His wife could not have more existence than she does in his mind. Nothing in her touch could equal the joy he now feels in her presence. He thinks that even if someone magically delivered a message to him at this moment as he trudges along in the wind and this message told him of his wife's death, this would not alter his joy. He would continue to see her image in his mind, continue his conversation with her.

Love would keep him alive and love would keep her alive even as she moldered in her cold grave. Frankl now lives inside his mind. He enters warm apartments, he answers the phone, he turns on lights.

He cannot be broken or destroyed.

O'Shay stares down at the man and wants to join him and tell him to think not of his wife for a moment but to talk to Joey O'Shay.

HE STANDS WITH HIS BRUSH and thinks.

He has most of it: the night, the hill, the explosions like flowers erupting from the valley below. The man is on the hill watching in the long night of the war. He needs one more element, the sky that hangs over everything and watches and does nothing as the wounds continue to bloom below. This part O'Shay is sure of, confidence fills his hand as he picks up the brush again.

He paints the sky and the stars, yes, the stars wheeling everywhere, flickering from their own internal fires. This part of the painting O'Shay knows in his bones.

O'Shay never wonders about this painting, not why he paints, not why he painted this canvas. Or why he made the stars in the sky that way.

HE CANNOT SETTLE DOWN. Sonrisa is dead, it is later, and he cannot settle down. Killing a policeman is a violation of an unwritten rule and now this rule has been broken.

During the trial, O'Shay goes down to see the face of the killer. He is crestfallen when he does. The guy is a little nothing, and O'Shay does not like the fact that such a man could take out an Angel Sonrisa. He does not know what he expected or what he wants but he knows he is disappointed.

The man goes on death row and for more than a decade works through his appeals, goes on living and eating while Angel Sonrisa goes on being dead. O'Shay works on this theory he has about who dies and who does not. He sips coffee and says, "A lot of cops who have died had this special aura that said they were goddamn good people, they were happy. It's like they were ready, you know?"

His old partner had that aura, he was one of the good people, he had a happy glow about him.

Then there was another guy, a Native American, and he had that glow, too. He pulled over a car to tell them to turn their headlights on and was cut down instantly.

He knows he is not one of them, the good people who have this glow.

That's why the killings cannot be discussed. That's why the war stories are ash in the mouth. They are fantasies we use to avoid the real stuff, they are the bullshit we call fiction.

Look, he silently tells himself yet again, you can say life is just biology, just a bunch of juices and drives and it goes on and on and it doesn't mean anything, just this big fucking wheel of pain. But he cannot abide that, there has to be more.

"I ain't no angel," he says, and he is speaking very, very softly now. "If life is a learning process, maybe the ones who are angels have found the answers and those of us who aren't have more to suffer still."

He's got that ice pick, his gun is always at the ready, he's covered, by God. Except for this one deep flaw: he insists that what he is seeing and tasting and living must have meaning. This desire on his part is what almost tears him apart at times. And this desire on his part is the reason he has kept going year after year after year and not fallen apart. There may be only two kinds of people, those whose lives have meaning and those who insist life is meaningless. O'Shay is determined to find some niche in the first group.

But he cannot conceive of an acceptable meaning based on what he sees and what he does. He hurts and cages people and sleeps well afterward. His boy is dead and nothing makes that right. He drinks in the night and remains thirsty. His friends are slaughtered and he knows he has become something the word *hard* does not accurately describe. He has become mean and hands out punishment without cause or justice or regret.

He cannot apologize for his behavior.

He cannot seem to stop his behavior.

There is a custom of action and then repentance. He cannot repent, any more than he can tolerate his behavior being condoned or ignored. The last thing O'Shay wants is forgiveness. He wants to understand.

IRMA INSISTS ON HER VISIT and O'Shay fits her into his schedule. Gloria is busy arranging for the heroin shipment, her almost daily calls keep him up-to-date and also keep him in her mind. She is part of a second feature, O'Shay tells himself, Irma is a lead in the main movie, the big heroin deal that will prove he can beat Colombians at their own game.

The room is like an old friend, he has done so many deals in this room at Bobbie's hotel. As usual, he's made sure her room was wired and he'd taken a room himself and it, too, had been wired. Pinhole cameras stare out. But now Irma is drunk and she is sitting on the bed, about to fall out of her dress and Christ, he can hear the hum of the air conditioner in the room and feel the eye of the tiny camera watching from above. He leans forward in his chair at the edge of the bed, and Irma is waving her arms, explaining, explaining how she will put the shipments together, how her connections in Colombia are golden, and indeed they are, since the wiretaps have revealed her to be the sister-in-law of a cartel leader, and O'Shay listens, his face a blank, and then says, "Yes, yes, but you've got to understand that we must do this as a business," and Irma pauses, and then goes back to her drunken babble.

It had seemed like a much simpler day when Irma arrived at the airport. He waited outside in his limousine and sent two of his men in to escort her and her bags. The car had a bar and different-colored lights that could be varied from a console, and she played with this equipment as the city slid past the dark-tinted windows. She got off the plane and O'Shay saw she was decked out in every damn kind of gold and emerald thing he could imagine, her skin light, her teeth perfect, her body a rock shaped by surgeons.

He told her, "Well, tonight we are going to go to a really nice restaurant, and you've been so interested in my airplane business, after that I will take you to my hangars."

When she got to her room, she called Colombia and said, Hey, these people are solid, they had a driver pick me up, they put me in a good hotel, these guys are mafia, the old guy, Joe, owns this hotel.

At dinner, at a restaurant that hangs on a cliff by downtown, they talked prices and about what O'Shay could do and what Irma could do. They touched on all the possible benefits of doing business together and she told him, We will do this, ten kilos of heroin at a time, the connections will be there for us. And so they sat talking about millions of dollars worth of heroin like a shipment of hams.

They had steaks, and below them the trains ran like toys. He noticed Irma kept drinking and drinking. And laughing loudly with Matteo.

Afterward, Irma simply wanted to drive around in the limo and see the skyline and have drinks in the back. And so they headed out to the airport and Irma started throwing down cognacs and by the time they got to the hangar she was drunk.

He took her in and showed her his fleet and she was excited by the large, expensive planes. She took photographs and beamed at the polished forms. And then they stepped outside and the night air washed over her and suddenly with a squeal she threw her camera into the night. O'Shay sent one of his men after it, realizing that he had a problem. She was a major person and he was responsible for her and he could not afford any stories of some untoward moment that occurred in his city while she was under his protection.

Back in the car she was suddenly on the phone to New York with one of her men, and she talked about the limo and the fine time she was having and O'Shay knew that whoever was listening had to know she was drunk, very drunk.

He could hear her saying, "No, no, I'm not drunk," and then there was a slight pause, and she insisted more adamantly that she was not drunk.

Then she handed the phone to O'Shay, and a voice said, "I would hope that you will respect who she is, because she is very drunk."

"Listen, dude," O'Shay replied, "I know she is drunk but let me explain something to you. We are showing her the utmost respect. I am taking her back to her room now. She is really drunk and we are not going to do anything to her and when I get her to her room I will call you then. But please understand this, she had my limo driver get her a bottle of cognac without my knowledge or consent."

And the voice broke in, "I understand, but normally she eats very little and drinks very little and so when she does drink... she goes insane. She is spoiled but please, remember who she is."

O'Shay hung up and Irma smiled at him and said, "Fuck him," and turned the music very loud. And she started squirming in her seat and dancing to the music. Then she started taking her clothes off and O'Shay thought, Oh God, no. She was unbuttoning her blouse and O'Shay could sense that in a few seconds she would be displaying her expensive breast surgery in his face.

Then she began breathing deeply, and he could tell she was about to vomit.

They pulled into Bobbie's place and Matteo and Angelo took her into the hotel. Irma made it to her room before she buckled and retched.

When they entered the room Irma couldn't find the fucking lights at first and she fumbled around and then tried to turn on the radio and all she got was some kind of piano sonata, not really classical, that softer stuff made in some mill to calm people with money so that they can feel safe with their money and this bland music irritated her and she snapped, "Chingada!" Fuck! Like some street Mexican. Eventually she collapsed onto the bed, and they got some lights on and settled in for this brief conversation.

O'Shay's eyes wander the room as she talks, take in for the hundredth time the dark rosewood furniture, the desk, coffee table, armoire, low couch, the dark heavy headboard, the wall-to-wall carpet with the complex vomit pattern all hotels need and, from the way Irma is acting, a pattern this room may soon need. The cream walls, the delicate frieze of leaves, the bedroom, then the sitting room, the bathroom which has a Jacuzzi and large vanity, everything nice, not too much, but nice, just enough to say, You matter, you are under my care, you will be treated with courtesy. You will sit on the fucking bed dead drunk, your face glowing with the light off the large brass lamps, and you will babble.

And outside bums will prowl the parking lot, and downstairs in the lobby it will be serene and comforting and all the Irmas will feel safe, feel secure in this strange place in this strange city, secure enough to deal.

Matteo makes her get on the phone to New York and tell her man she is okay. And then she begins to calm. And sits on the bed and talks endlessly of jewelry.

The next day she is hungover, apologetic, and wonders where

she might have put her camera. O'Shay tells her they could not find it, but it hit one of his planes parked outside and dented the fuselage.

She is sick over this damage but he assures her it does not matter, that drunkenness does not matter, that only the business they will do together matters.

She ends her visit, flies back to New York. Her calls are traced and then tapped. Her visit is now video. Her new business is now a code-named operation, one that causes dreams of promotion for some in the agency. In Washington, the cases become a high priority, the ticket not only into the Colombian drug world but also to the higher reaches of the U.S. bureaucracy.

Gloria and Irma now talk to machines and are scrutinized by strangers. The people they call enter a file and in turn have their phones tracked. Already, Gloria has eight keys of heroin lined up and soon it will be more. But the government is not anxious for a buy, they are becoming addicted to listening into this new world and so they ask O'Shay to delay things.

This is not easy.

He is not living within agency time now, he is living street time and the street moves and works and insists and acts. The street does not shut down at night, the street toils through the weekends. The street actually delivers the goods.

POPSICLE'S MOVIE is not part of O'Shay's expectation. O'Shay goes into a back room, finds the ledgers, the records of a drug business. The younger guys on the raid are in the living room riffling the videotapes. They love to look at the tapes because they know so many dealers like to videotape fucking their bitches.

Suddenly one of the cops is at his side, and says, Hey, you have to come out here. He follows the guy out to the living room.

And there on the screen is a girl being strangled and intermittingly begging for her life.

O'Shay is enraged, in part because he knows the tape might not be admissible.

But also, he knows who the girl might be. Homicide has found a sixteen-year-old dead at the base of a seventy-foot cliff from where someone pitched her off.

He stands there and looks at Popsicle's movie.

THE FOG SMOTHERS THE LAND, the fog looks warm out the windshield but the fog makes the entire world seem cold and dead and forgotten. O'Shay sits, says nothing, the car rolls through the fog, dips down the hills to the river bottoms, then rises up the facing slope, lights bleeding against the thick, wet air, that glow that never really penetrates the murk, the electric fire that gives off not a hint of warmth, he lives mile after mile after mile in the fog.

He is not alone.

Cars roll by like ghosts and he knows they are all headed to the same place, that they all feel the same coldness as the fog softly licks the land.

O'Shay feels numb. But he cannot blame the fog for his numbness.

He knows that this dull lack of sensation is all that he can possibly feel, or all that he can risk feeling. He is not himself. Or, even worse, he may for these few hours he is entering actually become himself and no longer have the armor of his attitudes and his business to protect him.

Cars roll by, no one waves, but still they are part of a silent formation moving through the fog.

Atop each car a red light bar blazes and the color plays against the fog.

It seems like the highway is all one flow, and this flow heads to the burying ground to say goodbye to Angel Sonrisa.

THE DEALS MARCH ONWARD. Irma flies in with little warning and wants to talk business. She asks O'Shay if he can move a shipment of cocaine with his airplanes. O'Shay ponders this request.

O'Shay says, "Look, I can lease a Learjet. Or if time is not that important, I can move it with my own twin-engine plane."

"How much do you want for this service?"

"Well, look, if I'm moving cocaine, say 200 kilos, then I want $1,200 a kilo for the transport. Which you know is not a bad price. But for heroin, I want $5,000 a kilo."

"Yeah, that'll be good."

After three days, she leaves.

O'Shay is in control. His deals are now a priority in Washington, the phone he loaned to Garcia has produced a list of numbers the likes of which the agencies have never tasted before. He is juggling two major heroin operations out of Colombia, he is suitably stressed and preoccupied. But it is all flat to him and he knows it.

He knows too fucking much. He knows what Irma is saying to her people in Colombia and New York and Mexico. He knows her numbers per kilo better than she does. He knows also that Gloria is offering him heroin, 94 percent pure Colombian, at around $75,000 a kilo and that she is paying $58,000. And so he sits and haggles, in Irma's case for three days, and it is all bull-shit because he has reached into her life and her mind and can read all of her cards and she is blind to his hand.

The mannerisms, also, are no longer fresh and intriguing to him. He knows the scent of Irma before she enters the room, the smile, the fine fabric she will wear, the badge of gold jewelry, the surgically enhanced breasts her clothing will display, the trim body nurtured by bouts of liposuction, the dinners at fine restaurants, the idle banter that must go on before the moment comes to talk deals. He is like an absentee at his own case, a man technically present but barely awake.

He sits in his city and people in other nations go to work in order to supply him with heroin. His phone rings constantly—Irma, Gloria, Cosima, Alvarez, Bobbie, Garcia, Alma—and he shuffles them like separate trains rolling down his private set of tracks. He thinks it should not be this easy, there should not be a fucking kilo of heroin in the country if it is this easy to lure the suppliers into his world.

A part of him lives off the craft of it, off the ability to read other people and entice them and then betray them. His ability to destroy them. But for this to be fully satisfactory, they must meet certain conditions. They must be bright and wary and dangerous. They must be bad and even better than bad, they must exude a palpable evil. And his sense of these conditions is eroding. In his head he hears Garcia talking about his children, Gloria is leaning over to show a photograph of a grandchild, or her daughter is on the phone making conversation, or Irma is sitting there acting foolish and looking spectacular. He is flooded with impressions of the humdrum details of their lives, the utter normalcy of their needs and their loves and their little fears and vanities.

The sense of being on the right side of some line is getting harder to sustain.

People are going down because of O'Shay's hunger and his need to feed some part of himself and because of his ability to deceive and betray and somehow this does not feel as good as it should to him.

He has a big file cabinet full of videos, the rough footage of other people's lives. O'Shay shuffles the tapes in and out of a player in his office late at night. His fellow agents are all gone, the walls hang back in the semidarkness, the room is mainly the hum of the huge air-conditioning unit and on the little screen black-and-white figures talk about kilos, planes, jewelry, children, grandchildren, concerns, small voices with small notions, the dribs and drabs of humdrum life. Here is Alvarez sitting on a couch in the apartment, nicely dressed, flinging himself forward on his hands and knees like a dog on the carpet under Angelo's gun, his eyes almost blank as if some footage is flashing before them.

Garcia sits with O'Shay in the apartment and suddenly Garcia is demanding O'Shay's home address and O'Shay rises angry from the chair and he is not acting.

Irma sprawls drunk on the bed and babbles as she fights off the urge to vomit.

A table holds a hundred kilos, a busted dealer says "I can't believe you're a cop," a man lights a long cigar, exhales, and gives his shopping list of desired murders for O'Shay and his boys to do.

Gloria glows with that radiant smile, that girlish toss of the hair, the smooth skin of a child, the tenderness in her eyes and O'Shay sitting there still, all business, his face a lie. Cosima moves dishes, Angelo plunks down his bottle, the rattle of the package of chips, the flutelike voice of Gloria trills from off camera,

O'Shay's drawl, now and then laughter, Cosima and Angelo ignoring the talk like two waiters who do not interrupt their betters, all the decorum of a serious heroin transaction observed until suddenly, after twenty or thirty minutes, O'Shay and Gloria appear in front of the camera, and they look buoyant and alive.

He moves with almost a grin, his sport coat open, his shoulders relaxed. Gloria holds the strap of her purse across her shoulder like a sorority girl crossing the quad. And then amid a swirl of chatter they all leave the apartment, the door closes and there is no sound left but the hum and sighs of the refrigerator, the creaks and murmurs of the building itself, the tape keeps rolling, staring at the empty kitchenette and the bottle and food Angelo has left.

O'Shay clicks off the tape, remembers Cosima banging shit around in the kitchen, back in O'Shay's orbit now that she'd set up her friend but sidelined as a flunky while O'Shay and Gloria settled the outlines of the deal. This is the way things are done, but she was still pissed, and she rattled around so loudly that he was certain Gloria noticed something was off and so he said, Hey, let's go get something to eat.

He remembers now how it just gets worse. At the restaurant, a fish joint he likes, Cosima is rude to Gloria, to Matteo, to the waiters, to everyone.

Gloria tries to salvage the situation. She says, "Hey, I like this Cajun type of seafood." She has one glass of wine and remains cheerful.

Gloria adds, "I think I'm going to do this deal through connections on the Gulf, I can get eight keys Colombian there."

And O'Shay sees it all ending in a simple buy/bust and then the bracelets. Or maybe not. He wants the deal to keep rolling, for the wiretaps in Mexico and Colombia and other places to keep listening. And so he conceives of a scenario where Cosima and Gloria go and get the heroin, are stopped in their car under some police pretext, the heroin is seized but the two women are allowed to escape. Yes, he thinks, that will do it, keep the thing alive and yet under control.

As they walk from the restaurant, O'Shay notices Gloria's scent, and this time it carries a hint of ginger, yes, he thinks, that is it. He notices that she has lost weight since her last visit, and that this new sculpting accents her full hips and large breasts, notices all this while he invents the escape plan for her and Cosima. It's true what Cosima told him, that Gloria was dieting because she liked him, and now he can see the proof as she sways ahead of him.

Suddenly Gloria says, "I want to spend some time with you where we can just talk alone," and she hugs O'Shay.

"Look," O'Shay says, "if the weather warms up, we'll go out on my boat."

He will take her up to the lake in a luxury car, yes, that is what he will do, and put her on his fine twenty-two-foot boat. It will be the frosting on the cake.

But more than that. To go to the lake, for O'Shay, is to travel inside the place where his soul lives. That is why he must have the lake and also why he seems hardly ever to make it out to the lake. The lake is where he found the earth, where he walked in early morning down to the cove, slid the jon boat into the water, and floated, a boy in a green world. It is a place beyond the complaints and habits of adults, an actual place. To take her to the lake is like taking her inside of his heart.

He is himself, relaxing inside his own skin, and he is winning Gloria's affection because he feels affection for her. She is what he likes in life, dark and sweet and intelligent and full of curious habits like watching old movies. He is not faking anything, including the fact that every minute he spends with her absorbing the pleasure of her company, he is plotting her ruin.

They spend the day on the water, wander amid the islands,

trace the shoreline. Gloria sits and watches and laughs. On one island, the soil is almost white, the trees and shrubs whisper peace, and the air is so clean it seems to blow from Eden. They hardly talk and yet always communicate. The world has fallen away and there is nothing but the water, the green, the blue sky, the soil under their feet, the sanctuary that islands always offer, that sense of being safe, not simply from a corrupt world but from one's corrupt self.

A plane comes in low, and lands behind a mansion on a hill across the water.

"That's one of mine," O'Shay says offhandedly, "and that is one of my houses."

Gloria laughs.

"Two hundred pounds of weed," O'Shay continues.

He is violating one of his own rules here, by offering information, by making up something on the spot. He knows better than this, he knows such little snippets of claimed facts can trip one up. He is not being himself and yet he cannot resist the moment or the glow it kindles in Gloria's eyes.

On the way back to the city, O'Shay drives with Gloria in the front seat beside him, the music plays low, her skin has the warm glow of sunburn. He can barely stay awake, he is exhausted from being the grandee showing his girl the lake. And yet this fatigue is interlaced with a kind of excitement, a sense of having spent a perfect day in the place that always offers the possibility of healing him.

He tells himself that technically it has been a good move, that she now trusts him so completely she will miss things and be easier to herd toward that prison cell. She will lose her shrewd judgment, her natural caution, that essential alertness, and he will have an enormous advantage.

"You know," Gloria tells O'Shay, "I don't ever want to go back to the penitentiary. You know I did time, don't you?"

O'Shay nods and says, "I don't want you to go back either."

She looks over and asks, "And you?"

"Long ago, when I was young."

And they pause and O'Shay asks, "You're not married, are you?"

"No."

She hesitates and asks, "Aren't you divorced?"

"Yeah."

She relaxes, the car roars down the highway.

Gloria says, "This thing that we do, it is important. But it is a thing that is hard to get out of."

She wants him to go with her to the Dominican Republic, she says softly. She has two houses there, one on the beach. It will be like their time on the lake, only better.

O'Shay interrupts, "You know I have an even larger boat, it was at the dock when we came in, but the one we used is faster and in many ways more fun and..."

"You need to come to New York," Gloria says, "and meet my kids. I'll show you the city, it is fun."

"Well, I don't know," offers O'Shay, "it's a big town, it may be too much for me."

"Oh, no, no, no," she says. "I'll teach you about Dominican food."

And then she leans up against him and falls asleep.

She is safe now.

And O'Shay knows from all the overheard conversations that New York carefully transcribes that Gloria is not having affairs, she is having nothing to do with men except business. He knows from Cosima and from the intercepts that Gloria has not been with a man for at least five years. She is saving herself for Joey.

She tells him things like, "This is business between us, but I like you. And Cosima is jealous."

"Yeah," he drawls, "she's fucking crazy at times."

And now they are eating in another restaurant and Cosima is acting out again, but O'Shay has stopped worrying about this behavior since Gloria has filed it away in some dossier labeled "jealousy." O'Shay drives Gloria back to her quarters. They go over and over the deal.

O'Shay says, "Don't worry, don't worry, I'm not in a hurry."

"But I am," Gloria breaks in. "When I tell someone I can do something, I can. I'll show you."

"Look, I like you enough, I don't even care if you deal dope."

"Oh, Joe, I don't have to do anything really to be happy. My idea of a good time is to lie in my bed and watch old movies. And that is what I do."

They pull into the hotel, another place managed by Bobbie. It has microphones and cameras at the ready in a room O'Shay has created for business meetings. But not in Gloria's room.

They ride up in the elevator, and they are quiet now.

She opens the door to her suite and motions him in.

She closes the door, removes her blouse, and her breasts flow out of her brassiere.

O'Shay hesitates.

Gloria kisses him and says—in English for the first time, in perfect English—"I want you, I want you." She is pulling her pants down, dropping her bra to the floor. "I want you now, I haven't been with a man in years, I need you." O'Shay is paralyzed.

"Look," he sputters, "this is no good. This will get in the way of business. But as soon as the business is over, I want to get with you. I want you really bad."

And she sighs.

"I want you too. Maybe you are right."

When he gets down the hotel hall, his men ask, "What happened? Something happened, right?"

"Yeah, she started undressing."

Matteo asks, "Were her tits big?"

O'Shay says, "Her tits were so big I almost fainted," and they all laugh.

The next day, he and his crew take her to the airport hangar. The businesswoman is back.

She looks at his planes and says, "Okay, this is perfect. If we can get things to here, to you, can you fly them to New York?"

"Yes, I can do that."

He can see her eyes sweep the hangar, calculating loads and flights and deliveries. How they can move things from Colombia to Mexico and from Mexico to here and from here to wherever the loads are desired. He can hear these thoughts tumbling in her mind as she itemizes his fleet and his potential for her.

They step outside the hangar. A cold front has moved through and dropped the temperature down into the fifties.

She turns to him and says, "You know I love you."

O'Shay nods, says nothing.

He forms another plan in his head. He will tell New York that when they take her down, they should let her plead to something because he is determined to continue hunting major Mexicans and Colombians and needs for her to slip quietly into the shadows and not alert them to his true identity, something a trial with the evidentiary rights of the defense would almost certainly reveal. Yes, that's what he will do.

Besides, he tells himself, if she does not go down, she will be murdered for costing her suppliers so great a loss of goods and money.

And he does just that. He calls the agencies and he says, "Look, here is the deal. You are risking my life and my informant's life if you expose us in a trial, and you will end this investigation."

They hesitate.

O'Shay tells them, "Look, I don't know if I can continue to do this if you don't agree, I don't know if I can live with the risk or let my people be exposed to such a risk."

As he says these words he knows they are an absolute lie. But the lie works. They agree to let her take a lesser fall if she pleads. She will not go down for the full weight of their deals, she will not die in a penitentiary serving out life without parole.

He has designed another escape hatch.

Gloria is back in New York. She is working the phones very hard, he gets reports from the agencies on her toils. O'Shay knows she is doing this because she is in love with him. She's seen so much in life, he thinks, of the vicious bastards in the business, that maybe he appears to her as hope, maybe for something beyond a life of total deception, of getting fucked over or just fucked and left.

He calls her constantly—"What are you doing? How's my baby?"—and he knows he is using her. He doesn't push about business, no pressure on his baby about making those promised deliveries because the agencies are devouring calls in Mexico and Colombia between major people about major shipments and they want the lines kept open and Gloria to keep talking.

Gloria's children talk easily to O'Shay, sometimes they interpret for their mother so that she can be sure her true feelings are communicated. And they are not in the business, they are her special creation, something clean she has made from something dirty. Of course, they know what their mother does for a living, but they are allowed to live outside this knowing in a nice house without any reminders of her work except the constantly ringing phones.

They greet him like an old friend now, "Hey! Joe! When you going to come and see us?"

The web expands and deepens, the satellites bite into conversations, the tapes roll, the agencies in Washington, New York, and cities of other nations feast on the flow of Gloria's devotion. Business is humming. Bobbie is running constantly, arranging limos, taking visitors to jazz clubs and restaurants, letting O'Shay run yards and yards of wire through the hotel. Sick as she is, O'Shay thinks the work is making her feel alive again. At times, when he looks at the outline of the case he has constructed out of thin air, he becomes excited, and as New York and Washington grow ever more interested, he absorbs their interest. From a case viewpoint, it is all under control.

And yet he makes notes to himself:

Will put Joe the dope dealer away for now. Joe makes me a little ill. Slept 15 straight hours. A bit groggy still. Had to make taped undercover/negotiations calls of Dominican side of this shit. Feel like I'm wading through quicksand. Cosima causing irritating and selfish problems. Closing in on the end though...I can see the light at the end of the tunnel and I believe there is fresh air there. I know I must break away... go some place where I can see the stars...maybe I'll make sense then.

THE BOOK IS VERY THIN and feels strange in Joey O'Shay's large hands. But he clings to the book because it convinces him that his pain has some purpose and that this purpose is not because of a plan but because of an opportunity. That he can make something out of this pain and this something can be good rather than bad. It is up to him.

He reads and rereads. Viktor Frankl says, "Even though conditions such as lack of sleep, insufficient food and various mental stresses may suggest that the inmates were bound to react in certain ways, in the final analysis it becomes clear that the sort of person the prisoner becomes was the result of an inner decision, and not the result of camp influences alone. Fundamentally, therefore, any man can, even under such circumstances, decide what shall become of him—mentally and spiritually. He may retain his human dignity even in a concentration camp.

Dostoyevsky said once, 'There is only one thing that I dread: not to be worthy of my sufferings.'"

O'SHAY CAN taste evil.

Evil is that thing when we were kids we knew was there...but as adults logic makes us think or hope it's really not. It is the grotesque thing within us we know we are capable of...a pure and absolute wrong of infinite dimension. We have all felt it...it's the dank smell from below...luring us away. For the most part we shy away from its magnetism...at least most.

Then there is the other "thing" within us that as kids we also knew was there, which is even more frightening...but as adults "logic" makes us think is sheer nonsense. The notion of absolute divinity and good. We know it's there, too...but we shy away from it...it is beautiful beyond dimension. We've all felt it. It's the warmth from above...luring us away.

For the most part, evil and goodness both horrify us...so we find some easy middle ground and deny that either exists. Not too many Hitlers...even less Christs. Whatever the soul is, it knows and it has nothing to do with logic. Our soul whispers to us that there is a loving something, God. Evil is the absolute turning away and betraying of that soft whisper...

HER PALETTE TODAY IS GRAY, all the shades of gray, with faint hints of white here and there. The city spreads on the lower part of the canvas, all the towers and spires, the Chrysler Building reaching up to one side, all the might and money suggested and yet rendered in the tones of a ghost. Imposed over this skyline and deathly sky is a blue outline of the map of the United States. The twin towers are gone, snatched from the innocent and dumb city that once hosted them.

In their place is something beyond the imagination of the stupid gringos: the massive cone of a volcano and from the caldera spew clouds of ash and heat. The real world has erupted in this city, exploded out of the sleeping ground and shouted to all the fools, lulled by their sheeplike sense of safety, the true nature of the world.

Cosima refuses to use bright colors today. She goes for the sensation that comes when you close a deal, when the heat of the moment meets the icelike sensation of doing other people.

O'SHAY DRAGS THE DESTRUCTION of Gloria along like an anchor as he moves through the city making deals. And some parts of his past make this work seem like easy work without any personal cost.

There is the time he goes to flush drugs from a high school. Some of the kids who use also sell in order to pay for their pleasure. The cops take the easy road of busting the kids although this accomplishes nothing, since the habits persist and the people supplying continue to deliver the goods. O'Shay wants to plunge the knife more deeply into the tissue.

He has developed this notion of the unknowing informant, the person who supplies information, who sells out others while remaining unaware of this betrayal. He selects two women for the case who look young and fresh. He tells them to enter the student body and target the kids who are loners, losers, who are most vulnerable to others because of their deep need for acceptance.

One of these loners is a boy. He is easy and soon he is getting them LSD and cocaine and then heroin. He ratchets up and introduces them to his suppliers. It is all humming along on schedule. They make 165 buys and deal with 113 people, almost all of them adults. The boy is happy, he has two new girlfriends, he is showing off to them, letting them see how connected he is.

When they arrest him, he is suddenly facing felonies. He tells the two young women who had posed as high-school girls that he does not blame them. He comes from a broken family and has been abused. He cries. And when he makes bail he overdoses on heroin and dies.

The second loser is a girl and she does not overdose. When the arrests come down, she realizes that she has been used by people she thought were her friends.

So she hangs herself.

O'Shay has orchestrated this drama off camera, directing the action, coaching the players. He has proven to himself the genius and power of this concept, the unknowing informant.

10

IRMA INSISTS ON YET one more visit. She says her main distributor in New York, a major man, must meet Joey O'Shay. It is essential, she continues. O'Shay pauses. He knows who she is referring to, since he knows who she talks to and what she says. The man is a fifty-year-old streetwise Dominican with businesses. There is no choice. He flies them to his city.

He has a limo pick them up, Bobbie takes them to clubs. He treats them right.

The next day, O'Shay shows up at the hotel. Irma has a room, as does her colleague, Victor. Alma is brought in from the East Coast and she also has a room. O'Shay himself takes a room so he can rest between negotiations. Down the hall, the hotel has a conference room called the library and that is also taken so that meals can be catered in for everyone. And all the rooms are wired.

Victor meets with him first and they talk about nothing—soccer, New York, the weather. Then he gets to the point. He tells O'Shay, "She is embarrassed about that time she came, about her drunkenness and she is embarrassed to even come in here and meet with you."

"She does not need to be embarrassed. Everyone gets drunk

at times. But I hope you know we respected her. She is an important person to us."

As he speaks, O'Shay takes in Victor, a balding man of about five nine, light-skinned, maybe 195 pounds. He wears a jacket, slacks, no jewelry. He is very polite.

"Look, Victor," O'Shay says, "there is a lot of money to be made here. I wouldn't be here now—and I've already spent a lot of money—except that I understand you are the main man that moves the shit. Let's have some dinner in a while and talk about this shit. There is a lot of money to be made in what I am already doing with hydroponically grown marijuana. It goes for three or four thousand dollars a pound and you don't catch much heat selling it. You know I could bring three or four hundred pounds to New York and you would get much more for it."

"Yes," Victor breaks in, "that is true."

"Here's what I want to do. I want to bring stuff up there in my planes, maybe four hundred pounds at a time, and I will front it to you. What I want from you is heroin for the flight back."

Victor listens and then says, "I am going to go back and see what I can do. It won't take long."

Then Irma enters the room. She still seems ill at ease and starts talking about her family ties to the Colombians and how she is divorced from a cartel leader and is raising their children. As she speaks, Victor nods his assent.

Then Irma says, "Joe, my main thing is not heroin. I need to move 6,000 kilos of cocaine."

O'Shay nods.

"I must get it to New York—that is my responsibility."

"Look, I can probably get rid of as much as 100 keys at a time, maybe 200, but I can't sell the amounts you deal. That's not really my thing."

"No," Victor says, "you don't understand. She needs to transport the cocaine."

"Oh, well, maybe. But I can't move that much in one shipment. I can do it in separate loads."

"Well, you understand your responsibility if you do it?"

"That's why I'm not going to do it all at once, Victor. I can move maybe 1,200 pounds at a time. Let me think about this and about what I will want. I might even take some of it as payment."

Victor nods.

It goes on for three days with food and talk and haggling, and tapes ceaselessly rolling. Light falls through the windows, voices softly rise to the vaulted ceiling, there are breaks for meals in the library and O'Shay keeps improvising, working over numbers, offering up prices, engineering the deal, and buying time.

At one point, Victor and Irma try to get a few kilos, a token really, shipped in from another city as a show of good faith to O'Shay. And this stretches O'Shay like a rubber band. If they bring the kilos, he knows his local office will insist on taking them down. New York is already anxious to take down everyone he has touched. But Colombia wants to keep listening to those delicious phone calls. Then the delivery gets stalled and O'Shay does not have to deal with this problem.

After three days, they leave with the outline of their business relationship settled.

Victor calls from New York and asks that O'Shay come up as soon as possible. He hesitates. He is the man, not a messenger boy.

Victor continues on, "Look, I can get you four kilos of heroin, I know that is not much but still, it is a beginning."

O'Shay replies, "Look, I'll send my men up in a plane."

And Victor agrees.

O'Shay can finally feel that some kind of end is near at hand. That months of stancing and enticing and arguing and drinking are coalescing into the crystalline moment prosecutors love: the buy/bust where the dope and the money are both on the table, silently recorded by the camera.

His men go to a modest hotel in Manhattan. Irma shows up, to their surprise, since such a moment should be beneath a person of her stature. Alma is there, too. The woman talk excitedly, the men mill around waiting for the heroin and the money to cross paths and for it to be over, for this moment of jeopardy to pass. Irma seems almost jubilant, she is on a roll and can feel the future cascades of money that will flow from this tiny beginning. Under her coat, she has on skin-tight jeans and, as a gesture toward a blouse, a French brassiere. She has come to celebrate.

They sit and talk and then there is a knock and one of Irma's pistoleros is at the hotel door with the drugs.

The bust goes down.

Irma screams, "Oh, my children, oh, my children, Victor! My life is ending. Oh, Victor, my God, what have we done? What have we done?"

Irma immediately begins cooperating. A DEA agent stationed in Colombia and Colombian agents fly up to interview her. She never guesses O'Shay is a cop. Victor says he will do his time and die in prison because otherwise he will be killed and so will his people. Garcia is in Colombia. They keep listening in on him and finally, they take Garcia down. He weeps, he bids his children goodbye.

O'Shay barely keeps up with these details of their ruin. He does not care. He did his thing, he got them into it, and he knows if he worried about what happens to the people he ruins, well, he would go mad. Nor can he afford to worry about what happens to the cases he makes when others take them over. The only satisfaction he gets is that the Colombian narcs are intrigued. One officer, who, according to legend, has killed around 300 people, wants to come to the United States to meet this strange white man.

O'Shay has pulled off the case of his career. He knows the case will be credited to others, and he likes this fact, and that it will cause promotions he will neither share in nor want. He is aware of this part of himself, of his desire to stay in the background, to be someone out on the street. The case gives him nothing, nor does his career.

It is after midnight and he sits down to make yet another note to himself. He writes:

Well...the deal went down in New York like clockwork. I had arranged to get keys of Colombian smack from Irma. After that week of intense negotiation here at Bobbie's hotel and wired up rooms...they had agreed to sell three keys at $70,000 each. Matteo and Angelo would fly a nice twin engine to New York with $210,000 for the three keys. They went... Uncle Joe stayed here...as any main distributor would. The head pricks here doubted...New York doubted...Colombia wire gods asked me not to deal so they could hear it some more down there...but I knew the time was now.

He pauses, sips some more of his drink. He continues:

They came, right on time in the hotel with the shit as I predicted. The cartel bitch screaming No Matteo! No Angelo! The Dominican distributor couldn't believe it...Fuck it. My guys call me. Ask what the Colombians will do to the families of the ones I've fooled.

O'Shay stops again. He tells himself he often keeps his presence out of the takedown so that he is never made, so that he can live to hunt again with no past trailing him. This makes perfect sense, almost too much sense. He takes another sip.

I tell them they will be violated and killed...worse... blamed...New York that doubted me now goes to drink beers...they call me and drunkenly pay tribute to the old gunfighter, as they call me...my guys know though and are sad...with the cruelty of the lies we fooled them with... perfect, learned lesson of the past lies...where ghosts whisper. In reality they are no more grotesque than a doctor over-prescribing some drugs...and they will likely die in prison and their families will be punished for letting an old cop trick them in an intricate, filthy, disgusting maze of false family and sham honor. I tell the guys it is a very cruel business we are in. They'll be home tomorrow. I'm drunk.

He pauses yet again, he needs more to drink. He knows there will not be enough drink for him tonight.

I sit and look into the fire of the fireplace. It's cold here tonight. I take no victory. I have more respect for the drug dealers I took down than the majority of the bureaucracy I work around. Tonight I will drink enough to numb the fact I have destroyed some other humans and most likely their innocent families and tomorrow I'll be sad.

He has soft music playing, the kind that is not so much composed as assembled, the kind no one can dance to because it disarms not only the ability but the desire to move, the kind that functions like a drug.

But there is one more thing he must write.

The other half of this thing still is to be finished...it will probably even be sadder. And the fucking beat goes on.

THE SONG IS ON A CD and his son has written across it a simple title: FOR MR. BADGE. O'Shay tries not to think about the title. Once the singer was a boy who got up very early with his little brother and raced into the yard to see if presents had mysteriously appeared in the knothole of a tree while they slept. The younger son is now playing college football and the songwriter is all grown up, and has fought a war in another country, and is ready to talk to the ghost of his father through a song.

> Come drive the back roads of my life
> The streets are dark and cold as ice
> The rain can't wash away the pain
> And the memories drive a man insane…in time

O'Shay listens over and over to the song. But he says nothing about the lyrics, he simply absorbs them.

O'SHAY TRIES TO REMEMBER when he was innocent and his work was innocent.

For about three years, he lived by kicking in doors with his old crew. They would meet in the office and leave with their guns, hit twenty to twenty-five addresses a month. They'd argue for hours about whether it was the first guy through the door who most risks eating a bullet or the second, but hundreds of times he was the entry guy, those firefights, and not a goddamn nick. The shouts, the screams, the moves toward weapons, the people on the floor, the alertness that he cannot even describe to himself, entering this plane of existence where almost nothing matters except this sensation of being totally alive because in an instant you may be totally dead.

Always because a Jimbo has betrayed someone and you trade on that betrayal in order to sell out the person yourself.

Jimbo is small, a very small man who all but bounces off the walls with wild energy. He lives for one thing, selling people out, and so he works for Joey O'Shay. Betrayal is like nectar for Jimbo and O'Shay marvels at how he deals with delight in destroying the people they know.

Jimbo leads him to another biker, someone Jimbo feels tight with and trusts. He likes it better when he sells out people he deeply trusts. He enjoys the thought of them finding out later who did them.

The other biker lives in a cheap complex where Jimbo claims the biker sells marijuana and speed.

The knock at the door is answered by a teenage girl. It is night outside and night inside with the front room all black lights, a booming stereo, and incense. People are sprawled everywhere smoking pot. Clouds float through the room. The girl leads them to a back room and O'Shay sees people piled up like cordwood in the rooms he passes. The entire place is a large, placid cell of music, incense, and mellow people.

The girl pauses and taps delicately on a door. When it swings open a huge biker stands there buffed from his stretch in the joint. He is at least six four and all muscle. Plus, a wide smile. Jimbo does the introductions with meth-fueled speed and the biker, who is drinking beer, continues smiling. His arms and hands are jailhouse tattoos.

O'Shay catches a movement behind him, a woman in her thirties, and the biker nods and says, "My old lady."

A .357 Magnum rests on the dresser.

They quickly talk price and the biker says in a friendly way that he can get very good marijuana and speed.

O'Shay nods and moves to depart.

The big biker says, Ah, come back and we can drink some beer and smoke some stuff.

It is all complete, the big biker, his wife, his daughter, the house full of incense and stoned people, music, that friendly smile, yes, it is all complete, a world safe and sane and happy.

The biker escorts them out through his warren of scent and sound, his old lady in tight jeans and a sweatshirt, him wearing jeans and a vest. O'Shay steps into the cool night air and turns and sees them smiling at him, the happy family.

The big biker had gone up on aggravated assault with a

deadly weapon, also on armed robbery. He had already given substantial chunks of his life to the prison system.

When they do the roundup, the biker makes no resistance. He never fights his case but pleads guilty and takes his ten years without complaint.

Jimbo, when they leave that night in the car, seems almost ecstatic.

He says, "See, see, he trusts me, he trusts me."

O'SHAY TAKES THE YOUNG GIRL to the office. She is a girl they have found in the apartment with a dealer named Q. And that video-tape of a girl named Popsicle.

He walks the girl by his voodoo altar to the interrogation room and she tells him her confession.

"I met Q when I was about seven years old. He was a friend of my mother's. I began having sex with him when I was thirteen years old. He would take me on dates, but people would think he was my dad. I am eighteen now. We have been having sex since I was thirteen years old. My child is his daughter. I have never been with another man."

The man known as Q sells cocaine powder to white people and crack to black people. She helps him cook the crack, she helps him count the cash, often a pile that tops a hundred grand. In the raid, everyone goes down.

O'Shay wants Q very badly for the murder of Popsicle. But the voice on the page of the voluntary statement, the eighteen-year-old who began sliding under Q's body when she was thirteen years old, will not give up the father of her child. So she goes down hard, twenty-eight years. And the movie goes into a drawer, a case that cannot be closed.

O'Shay watches the tape once. He notices the odd movement in and out of the camera. He can feel the operator is a female.

He can taste that she is a girl who started fucking Q when she was thirteen, had his child, and now at eighteen has been arrested with him. A girl who once knew another girl named Popsicle and operated a camera for about thirty minutes one evening.

GLORIA CALLS AND SAYS, "Joe, I can get you ten or eleven kilos. Come up. Or send Cosima."

O'Shay hesitates. Irma and Victor are sitting in cages but Gloria continues to operate as O'Shay delays and delays her arrest. He can of course claim policy reasons for this decision—keep those wires humming, keep that information flowing. No one really questions him about it. A week goes by, then two weeks, then some more weeks and still he hesitates to pull the plug on her life for he has, he realizes, become her life-support system. So long as he proclaims that she is useful, she is allowed to live.

She tells him they must get together as soon as this deal is over.

He agrees, yes, yes, once business is over they can be lovers.

Finally he deals with DEA and agrees to have Gloria taken down in a buy/bust. Cosima will be sent to New York to finish the transaction in the apartment Gloria keeps as a stash house. Gloria's people, being all business, being alert, being hardworking, and being in their own way honest, want to make delivery immediately.

DEA balks. Working a weekend is bad enough, but making a major heroin buy on Super Bowl Sunday? Unthinkable.

O'Shay says to himself, Those bastards are just flat-assed not going to do it on Sunday.

For two days, he's on the phone with Gloria, cajoling, making excuses and at the same time trying to prod agents in New York.

Finally, he tells her, "Just get me what you can get me by Saturday. Angelo has to get back now." He scrambles on the phone to keep the deal alive so that he can destroy her. "I've got a lot of money sitting on the ground, I've got a million fucking dollars sitting in an airplane in New Jersey and I gotta get it back."

"Well, Joe, okay. But what is one more day?"

O'Shay knows from the wiretaps that the Colombians are bringing it out of Rhode Island, the arrangements for the trip down to New York will cost some time, but he can't get the agents to wait, not even one more day for a much larger seizure.

Gloria is also anxious to get the sale over with. She has been shopping and has bought a new wardrobe. She has been to the hairstylist and gotten a perfect haircut. She has had a manicure. She is ready. As soon as the sale is complete, he is flying to New York, he will spend a week in her house, he will get to know her kids...and she will finally be in the arms of Joey O'Shay, the first man she has been with in half a decade. She's been running, been going to the gym, she has dropped weight, her body is firm. And yearning.

Cosima watches these changes and resents them. She tells O'Shay that Gloria has gotten too skinny. She tells Gloria that O'Shay is a terrible womanizer, that "Joe has all kinds of girl-friends, he especially likes black girls."

But none of this really registers, any more than O'Shay's floundering in his effort to explain the urgency of doing the deal a day early registers. Gloria trusts.

So Cosima goes to Gloria's stash house, she makes the buy, the agents hit the doors, and it is over. The bulk of the heroin is never delivered, Super Bowl Sunday rules. Gloria leaves in hand-cuffs having seen a drug deal gone bad and nothing more.

Her daughter, as soon as she learns of the arrest, calls Joey O'Shay seeking help with lawyers and bail. When Gloria discov-ers this fact, she is horrified. She tells her daughter to desist be-cause O'Shay is dangerous, and his mafia thugs will kill the daughter if she bothers them now.

For the first time in her life, Gloria decides to cooperate. Her only future, she thinks, unless she sells someone, is life inside a prison. The agents start walking her through her address book and asking about this number and name and that number and name.

Concerning Cosima, Gloria says, "She's just a friend of mine visiting, she has nothing to do with any of this."

As for the others, DEA realizes she is telling them the truth, something they can test with their endless transcripts from the wiretaps.

"Oh, this one," she says, "that's a major heroin distributor."

Finally, they get down the list to Joey O'Shay.

"Oh," she says, "that's my baby and I will never tell you about him."

They work down the list and then come back to O'Shay. They do this repeatedly.

She never gives him up, not that day, not the next week, not in the months that follow.

HE IS ABSOLUTELY ALONE NOW. He is in that distant city far from the cells and the sad eyes. He sits and makes another note.

For me, it ended today. My part of the "other half of the thing" is over. I convinced Gloria in New York that I was sending my pilot and one trusted family member, Angelo, with $748,000 via twin engine to get 11 kilos of Colombian heroin at $68,000 each. She agreed. Wires listening in New York, other cities, and Colombia. I stayed here and negotiated via telephone for two straight days as myself and one DEA agent fought the fucking bureaucrats that wanted to end it every evening and didn't want to work Saturday, but did... with a time cap. When Gloria had three (but would have got 8 more). She and her courier were busted while Cosima sat in the stash house with them. It's a miracle we got the

worthless fucks to even arrest them. Gloria told DEA that I
and Cosima had nothing to do with the heroin...She'll realize
it later...I suppose she'll be shocked and hurt.

Christ, he thinks, this is enough. He has cleaned it up, made
it into a simple trajectory of buy/bust. Of course, there is that
twinge, a little moment of sadness over Gloria.

But, he's definitely cleaned it up. Still, he can't seem to stop.

Wires on both cases are still producing kilos of heroin
here and in Colombia and more arrests. Twenty kilos of
Colombian heroin so far, Cosima responsible for another
20 Ks of Mexican heroin and a lot of organizations. Total
street value around 50 million dollars. Cosima is the best
CI I've ever seen...we parted company...she cried...will
be in NYC now working for DEA. Then she wants to work
Eastern Europe...probably won't live that long.

There is almost a telegraphic tone to the words, a faint click
click click in the lines as he sands down months, erases all the perfume from the air, blots out the stars, and makes it simple and safe.

It won't do.

He senses that.

He has to put down more.

As for me, Gloria was very pretty...polite...mannerly...
we spoke often and looked at photographs of her grand-
daughter. She spent a lot of time falling into the web...
believed I was a nice person (for god's sake) for a drug
dealer. She had that beautiful Creole skin...you know...the
African/white mix...it will turn yellow without light. She had
sad soft eyes...said she dealt to feed her kids etc. Her eyes
will hollow in prison. The whole thing makes me sick. She
will be very old when and if she ever gets out of prison and
so will all the others. It is all because of me. The intricate,

*filthy role I set up worked. They're up on all their fucking
wires and will probably "piggyback" (as the woofacrats that
wear nice suits and go home at 5 say) and catch other
fucking poor people trying to sell their illegal commodity.*

Closer, he thinks, a little closer.

But he's not sure just how much he can put on paper and
have stare back at him.

*Oh, well, I'll get Mexican food and I'll drink too much...
But it won't keep me from being disgusted. I hunted a lot of
humans and put them in cages for long ugly times...and I
wonder if it was all so absolutely...completely...horribly...
sadly...wrong.*

The next day Joey O'Shay returns to his note.

*It's hard. I also know it is very complex because I know it
is complicated with a lot going on.*

Ah, that's a little better, he thinks.

*So fuck it. If it is half what I feel, it will approach an old
topic in a poignant and beautiful way...where people have
depth and meaning has beauty even though truth is very
blurred. As for what follows me, others have seen it around
me and that in itself surprised me...and left them very
uneasy. They say that some evil guy dies and now sleeps...
let me assure you that such a person sleeps not at all. What
follows me...and others...that know...now has another
selfish toy to gnaw on.*

Near, very near, he tells himself. There is a limit to just how
near he can approach this thing. It is not simply how well he can
think or write. It is an absolute physical limit, a barrier that flesh
cannot overcome, a wall out there in space that nothing can

punch through. He can taste what is over the wall, in an odd way at times he can feel like he has passed the wall, reached out and found this other reality, but he knows this is a fleeting sensation, something that even as he touches it, he can feel slip through his fingers.

I've often thought there would be more people like Hitler and his close crew if given the opportunity. Though many are very foul in their limited domain. Think about this...just a wild guess...in the almost thirty years as a cop, I've probably come into contact with around 5,000 criminals and only a few were absolutely evil. I've probably been around about 5,000 cops/agents/detention officers in that same time frame...only a few were absolutely evil. One little thing I've noticed...evil must seek the right opportunity to ruin. To be a Christ, well, the opportunity is always present.

THE YOUNG WOMAN IS DYING and Viktor Frankl knows this fact and so does the young woman.

She says, "I am grateful that fate has hit me so hard. In my former life I was spoiled and did not take spiritual accomplishments seriously."

A tree grows outside the hut where she is dying.

She says, "This tree here is the only friend I have in my loneliness."

He looks up. All the girl can see is one branch of a chestnut tree with two blossoms.

She says, "I often talk to this tree."

He wonders if he is listening to delirium.

He asks her if the tree ever replies to her words.

She says yes.

She says, "It said to me, 'I am here—I am here—I am life, eternal life.'"

HER RED LIPS FLOAT ABOVE the tabletop in the hotel where Gloria once stayed. Cosima's mouth is not sensuous but pert, that small red schoolgirl bow. Her eyes are alarmingly clear, eyes that at first say nothing harsh in life has ever touched her and then, at a second reading, eyes that say nothing at all but exhibit an emptiness like a piece of glass. The face is very smooth also, a face with the skin of a child. That is Cosima's function, to reveal nothing.

She is showing copies of her paintings and in them the forces of evil seem to triumph at moments—towers fall, blood flows—but this evil is clearly labeled by her brush and denounced. O'Shay finds her artwork boring and is disappointed that others like it.

Cosima wears a black blouse, black jacket, and black skirt and this ensemble is stitched with red roses that match her lipstick. A gold chain around her neck holds a large medallion, a gold coin given her by her grandfather long ago, and her fingers gleam with rings she unconsciously twists as she speaks, or lightly taps on the tabletop when answering a question, so that her words have this background of click click click. She smiles easily but it is a confined smile, something that barely stretches her mouth.

She is laughing about the scandal that launched her into the life so long ago, when she ran off with the Colombian drug guys and her parents insisted she had been kidnapped.

O'Shay looks over at her, her face bathed in the soft light streaming through an east window of the lobby, a light with traces of darkness within like the light within cathedrals, and he says, "You were fourteen then, right?"

"No, no," she beams, at this moment her face is girlish, innocent, and delighted, the face of child relishing a new toy. "I was thirteen."

She explains how she destroyed Gloria.

Gloria was easy because she wanted money and so she did the deal for money.

Gloria was easy because she trusted Cosima, considered her a friend. And as Cosima says the word *friend,* that little smile briefly flutters across her smooth face.

She has a look of contentment as she reviews the whole affair. In prison, they lived a quiet life with those desserts she made, and lights-out each evening at ten. It was one of the most tranquil times in her life, Cosima says, a kind of vacation from years of fucking and sucking and having people killed and moving thousands of kilos, a holiday from work.

And now, she continues, Gloria will return to this calm, maybe for the rest of her life. She is delighted that Gloria does not even know that Cosima set her up, that she delivered her like a present to O'Shay and that she will make a lot of money, tens of thousands, for that delivery. She is delighted that Gloria will sit in her cell for years and think, Ah, Cosima, my friend.

Everything about the deal pleases Cosima.

The hotel is very still this afternoon. O'Shay sits silent as light trickles through the windows of the large lobby and Cosima shows off her artwork.

There is almost a feeling of love in the air.

When, years ago, Cosima was the lover of Carlos Lehder, the Colombian cocaine wizard, she fucked and loved him even though at parties he would often take a boy to bed instead of her. Victor once was a key aide to the fabled Pablo Escobar. None of the people she has ruined can imagine her motives, her relationship with Joey O'Shay, any more than they can understand who and what Joey O'Shay truly is.

She glows as she talks to O'Shay and he has a look of almost fondness as he listens to her.

Life has certain careless links. O'Shay brought Cosima back from Mexico for his work. She brought Gloria into his web because of her own needs. Now Gloria sits in prison because of hungers neither one wants to say out loud.

He gets up from the table, Cosima puts her artwork back in her folio, and they walk out of the hotel that Bobbie manages so well and into the sunlight that warms the city on this quiet Sunday afternoon.

HE TRIES TO TELL HIMSELF that he has been through this before. There have been moments of doubt, moments when the pain seemed too great, and yet he has always rallied and come back. Each time, he comes back because there is no place else for him to go.

He is in his thirties, unknown except to those who need to know. In an undercover world where a one- or two-year stint is more than most can bear, he has become the man who never leaves, who wears out partners like sets of tires. His name is not in the newspapers, he remains known to criminals only by his street name, Joey O'Shay, but in the dank and dark chambers where undercover people make moves and win trust and destroy those who get close to them, there he is known. And so they move him around, and ask him to come and train others in his unique skills. He is invited to another city to teach undercover work to a special unit of its police department.

How to do a drug deal. How to protect your life. How to manipulate an unknowing informant. How to break people to your will. How to be relaxed and yet ready to kill instantly, to kill without explanation or without even a nod to the rules because you are there alone and all you will ever be granted as an edge is that split second when you sense things going the wrong direction.

So as the class gathers and then proceeds, O'Shay notices this one young cop who is eager to enter and master the undercover world. O'Shay does not like his looks. He is polite, in good shape, and eager to learn. But that is not enough. He is simply too

decent, O'Shay decides. There is a fineness in him and this signals a lack of meanness, of the essential viciousness that will
keep a person alive out on the street. The kid comes up, wants to
talk and O'Shay cuts him dead. He tells him he will not teach
him anything, that he is simply not mean enough for the work
and should go home.

The kid refuses to accept this dismissal. He does go home,
he is out of the class, he does not get the benefit of O'Shay's hard
schooling.

Instead, he decides to prove O'Shay wrong.

He makes tentative moves, puts out feelers, and soon he is in
a genuine drug deal.

They take him to a cemetery, tie his hands behind his back,
and execute him.

O'Shay, when he learns this, tries to shrug just as he shrugs
at the fourteen-year-old girl who hung herself. Just as he shrugs
at Angel Sonrisa shouting, "Oh, shit."

But the shrugs cost him.

THIS IS NOT THE EASY TIME for Joey O'Shay.

He thought when the case ended, the sun would come out,
the creek would flow back through his life. He would sleep.

He would smell honeysuckle, the air would be thick with scent.

Sometimes on the phone, in those calls he endlessly taped,
Gloria can be heard blowing kisses.

Once she said, "Come to New York for my birthday."

*You can call it shame, shame of what I've done. I thought
I was having a nervous breakdown when it was over with.
There was a flood of guilt. I thought I had cancer. I've worked
since I was fourteen without taking time off, without feeling
ashamed.*

But now he wants time off.

One day he tries to stand up. And he cannot.

He sinks to the floor.

He crawls around.

He tries meditation. He puts on soft music, closes his eyes, and for a moment he goes to another place.

The place is empty but not simply empty. It is as if for a few moments he is innocent.

The snapping turtle did not design its tongue, nor did it design the fish and the fish moves near, the tongue waggles, the jaws snap, all this in the dark of a creek, on the bottom in the mud, where no one sees.

He stands up and retrieves his beer.

THE PHONE RINGS and he instinctively answers.

Joey O'Shay tends to business.

The voice on the line is a thug he encountered in the day, a white thug much like himself when he was coming up. And the voice on the phone beckons O'Shay, calls for him like some sound floating off a lake, signaling the wild.

He listens, sips some beer, and then barks, "Look, I know what you are, you're a gangster and you got to loaning people money because it paid good and now that's gotten you into something you don't want and so you want me to see what I can do about it. Am I right?"

There is a pause, he sips some more beer. The curtains in the house are white, and clean, white lace curtains.

"Okay, well, then I'm your man."

11

H E THINKS THE PEOPLE who sell drugs know what they are doing and so must face their time in cells.

For a year after the arrests he keeps working, mopping up other organizations. Back pages flutter across his life. When his dance with Colombian heroin was just beginning, O'Shay had also been dealing with a Russian who was offering Afghanistan heroin in exchange for fine cocaine. The night O'Shay took the Russian down in his fine mansion in the city, the Russian turned to him and said only two words: "You're dead." Then the Russian had been sold like a hog to U.S. intelligence, and his arrest record, the enormous assault on his mansion by a SWAT team and choppers, none of this surfaced in the press and the Russian might just as well never have existed. He simply vanished into the world of intelligence, a place O'Shay thinks of as the Dark Side.

Eventually the Russian wound up in Spain. Now O'Shay learns he has officially died of a stroke or perhaps a heart attack. He asks no questions and has no belief. When the local district attorney decides to finally seize the Russian's property, O'Shay leaves a simple message on the DA's answering machine: subpoena the Russian for notice of property seizure at C/O Lucifer, P.O. Box 666, Hell.

He can feel the Russian standing next to him in the mansion, feel the coldness coming off him, taste his calm as he suddenly discovers this is not a drug deal but an arrest. And he can think the man is evil and that he is a menace and that he must be destroyed. But he cannot escape a sense of kinship with the Russian, a feeling that he is a peer, a man who operates with his own standards and uses his own judgment. O'Shay is now the cop who feels an affinity for the robbers of the world.

He is taking down a major organization in the city and for three days and nights he has been managing at best five hours sleep. Faces, rooms, money, drugs, it is all a blur. He goes with his crew to a goat farm with a tiny brick house. They find ledgers full of business records, AK-47s, machine guns. And a cave of sorts, like the den of some beast with a huge altar and dried blood. Maybe the blood is from goats, O'Shay thinks. The Mexicans who work the farm are speechless with fear. They know of the altar and yet say they know nothing of its meaning or function. As the sun slowly rises O'Shay sees a photograph on the altar of a man holding a baby and the photograph is splattered with blood. At other locations, units find kilos of drugs and hundreds of thousands of dollars and fine cars and walls of paintings.

The sun is now up and he is talking to a blue-eyed Mexican arrested in the sweep. The man and his fat lover are both HIV positive. The blue-eyed Mexican freely confesses everything. He faces life without parole. He explains that he is talking because he wishes to do right by his wife and children. In Mexico, he says the special antinarcotics police trained by the U.S. agencies have now gone over to the drug world. Because they must eat, he says. And because the cause sucks. He looks into O'Shay's eyes and tells him that he knows O'Shay knows and what they both know is that they are fools living the life of fools.

What burns into O'Shay's mind is a tiny detail of the night. He and others hit one stash house ready for a firefight, guns out, nerves raw. As they break into the building a rat leaps from a pile of old clothes and scampers across O'Shay's shoulders. He is terrified, more frightened than he could possibly be if a man had emerged from the rag pile with a gun. When he finally gets home, he takes a long shower, takes it before he even opens that first beer. He can still feel the rat moving across his body.

And then a little while later he is in a different kind of home. The house is very large and the money is very clean. O'Shay has come here at the request of a lawyer friend to talk to the parents. The son is in rehab flushing his body of heroin. After that, he faces jail. The woman wears diamonds, the man wears the cares of business. The moldings in each room are twenty-four-karat gold, the kitchen is the size of a small house and granite counters race everywhere. The woman pushes a button and drawers open with various bottled waters carefully racked, selections of colas. There is the swimming pool. In the recreation room, an eight-by-twelve-foot screen smears images across a wall. O'Shay is uncomfortable. This is too much money. At night the boy lies in the room with the big screen and listens to music in the dark.

The mother offers O'Shay a drink.

He declines.

She shows him the boy's room and it is too perfect, too prissy.

He tells the parents that kicking heroin is not easy and they should consider failure as a possibility.

He does not feel comfortable. He is where the product ends, he is at the place Gloria told him about, the place of addicts that she did not create but serviced.

The woman is the perfect hostess.

When he gets home, he finds a message that two fellow officers have been shot in the face with shotgun pellets. They will live and have their own tour of rehab.

He does not pretend to fit these pieces of his day together. He simply slashes his mind with the jagged edges of the pieces.

He sits down and tries to explain himself to himself.
He writes:

*The larger controlling type/predator drug dealers never get
bored. They are in a ceaseless moil of brain strategy/planning
far far ahead of the average game. No cruelty intended but
for many dealers their major vulnerability is the fact they
get bored and only luck, magnitude of the drug trafficking
problem, and bad police investigative techniques protect
them.*

*To survive…like some major players do under constant
investigation…they…must make an intense relentless
dedication to never letting up…always watching for the
smallest nuance of danger or something wrong. They must
have multiple diversions and game plans and securities (that
can even be contemplation of who to murder that makes
them vulnerable). They can't let up. To catch these…you
must do the same.*

*There was always a battle for me…and more often it was
the quagmire whoofacracy that drove me nuts. Controlling
my frustration and severe anger at my own higher ups was
my enemy…not boredom. The mind strategy that it takes is
all day. People where I work would talk to me and I'd ignore
them. They often think me aloof or pompous or nuts. Most
people I ignore. They probably deem me weird or mean.
Most avoid me…some physically run off from me. I have
a reputation for cussing out prosecutors and have been
in trouble for physically threatening cops and more than
one prosecutor. They'd mill around wanting to know what
was next.*

*My mind was elsewhere…always trying to put my mind
in the opposition's mind and EMOTIONS (that more than
anything)…that is their vulnerability. The predators know the*

*cops' vulnerability. They know they wait in hotels...bored...
like the drones that sell/ move/ barter/negotiate for...their
dope.*

*I don't think I've felt boredom too much in my life...
perhaps sadness...in the stillness and aloneness. Much of
what I do is alone...usually now by choice...and in this
stillness away from the fucking strategy...it is haunting
regrets that chase me, not boredom. I am a half century
old now and the vast majority of my years has been in the
pursuit of what I have singled out as really bad "predators"...
only to sometimes find out they were so similar to myself in
some instances...people merely lost and wandering...filling
in the void with a vicious/violent game that takes them to the
edge. Perhaps in avoidance of the magnitude of life and the
finality of death.*

*In those alone solitudes...it is the self questioning and
need to know more about the meaning of life...that frightens
me. Almost a fear that when I stop pursuing them...I must
pursue who I really am or what I became.*

He pauses. But there is more, there has to be more. But he
knows that he will not get there because he is heading for a place
he does not want to know.

*I am struggling badly to do this "thing" I've done so long.
I feel adrift...and floating away from being a part of this shit.
I become convinced every day...more and more solidly...
it should be legalized...let the same bloated worthless...
overpaid empty-souled bumblers that are in charge of
enforcing the drug laws...dole it out...to the poor fuckers
that seek escape by means other than alcohol.*

HE KEEPS THE KID'S MUSIC around. He knows it is part tribute and
part rebuke. And he knows it is about him. He cannot feel sorry

for himself. He will not stoop to that. He cannot regret his life. He cannot shun his talents. And he cannot lie to himself either.

The night is fat with stars wheeling overhead, he looks up and takes them in. He thinks of the boy in the jon boat, he wonders what has become of the boy. He still has the boat, it rests by the water. But he knows he is not ready yet.

He punches in his son's music. He can feel the questioning in the words.

Still, he listens.

> Where bullets fly and Angels cry
> But they don't cry for me
> Where young men die and Angels fly
> But there's no peace for me

HE STANDS BEFORE AN AUDIENCE in a large meeting room of the hotel. He is wearing jeans and a sweatshirt, has a short beard and quick cop eyes. He says, "I have to step back in time and go over my life with you. I am involved in my last undercover case and there's a time when you realize it has to be over."

The audience is silent. They are parole officers. Bobbie is just down the hall in her office. She refuses to come in and listen. She does not believe a word he is saying or will say.

O'Shay goes back to how he learned to catch butterflies when he was four, how his mother taught him. He talks of the creek, the odors, the turtles.

The audience stares at him.

He is something unnerving to them. He is the slaughterhouse that murders the beasts they tend and process.

"My heart is not in it," he continues. His voice is soft, almost soothing. His face is blank except for the eyes. The eyes are pain. He has consented to talk this day because he thinks if he says some things out loud, then he can kill those things.

One of the things he wishes to kill is Joey O'Shay.

"Sometimes," he rolls on, "you realize you are so good at something that that fact is a danger. And I want to get out of here alive."

He does not explain the "here" that he wishes to escape from alive. No one in the audience seems to notice this fact. Or care.

He looks at them and says out of the blue, "To tell a kid that if he sells drugs he will not make money is a filthy lie. Alcohol is the drug of cops. It's legal, cheap. They drink that alcohol to obliterate the image of those people they have destroyed."

He sees that Bobbie is absent. He says, "Bobbie does not want all her bankers to know that some of the biggest heroin dealers in the country went down in her hotel."

The heads come up, puzzled. Bobbie? Who is Bobbie?

"I think," O'Shay tells them, "I learned to cuss before I learned how to spell."

He tells them that drug people are smarter than cops, and that they have a stronger work ethic.

He is speaking without notes. And he never stumbles. He is not speaking in a normal sense. He is reading some long message to himself. He speaks aloud to an audience of one.

ONCE, ON A TAKEDOWN, O'Shay had a stun grenade go off right under him. There was an enormous explosion, a huge flash of light. For at least ten seconds he felt out of his senses. His hearing never fully recovered.

He does not think of that moment often, or mention it, but when he does the explosion and flash are serene, a safe place rich in sensation and pure and clean and sealed off from the grime of the world.

The words are almost licked by his tongue and he sounds exactly the same way as when he describes being a boy in a jon boat at night floating in the quiet cove off a river, stars wheeling overhead in the moonless sky.

HE HAS GIVEN THEM the butterflies, the tree climbing, the lushness of the creek, he has told of being part of a gang, the life of the streets. He is patiently constructing the happenstance of his law enforcement career, the fact that he somehow went undercover and stayed there without coming up for air for twenty-two years. He is determined to lay it all out.

Except for the boy. There is no mention of the clean scent that is hardly a scent that comes off a child's hair.

Viktor Frankl writes that "of those aspects of human existence which may be circumscribed by: (1) pain; (2) guilt; and (3) death . . . How is it possible to say yes to life in spite of all that?"—well, O'Shay is not sure he should mention him either.

O'Shay looks at his audience.

"I have never found," he suddenly erupts, "one single person involved in drugs who was happy. Even the millionaires can't rest. The drug-dealer businessman has to worry about kidnapping, torture, and his kids becoming addicts."

He does not say it, of course, but he is just such a drug-dealer businessman.

"We are looking," he continues, "at a system that has absolutely failed and is an absolute lie."

He bites off the words like a shark tearing at prey.

Viktor Frankl says, "We must not try to simplify matters by saying that these men were angels and those were devils."

"The type of people," O'Shay states, "who sell drugs really aren't that different from us except that they sell something illegal."

Gloria is shown her address book with O'Shay's name and phone number.

She whispers, No, not him, he's my baby.

O'SHAY CRAWLS THOUGH the dark of the attic, the air close, dust floating into his nostrils. Down below, the cops are tearing the house apart, a compound that swallows up most of a block. They find a million in cash, faces, names. For years, Gordo has operated in plain view. He is a cousin of Johnny Boy, the serial killer with the dead eyes who found his calling as a savage on the street. Now Gordo has gone into the black box under the burning candles and skulls. There is a list and O'Shay thinks he will finish off that list.

He feels a biting sensation under his clothing and bullet-proof vest. A brown recluse spider has found prey. The surgeons will chop out a chunk of his flesh. Gordo will remain calm. He will go away forever and without a complaint.

In the neighborhood around the compound, people watch the raid. For days junkies pick over the huge building like carrion eaters. An old woman looks at the walled estate and clucks. She tells O'Shay in Spanish that she is happy Gordo has gone down. The house was filled with huge, garish portraits of Gordo and his women and children, tracing the lineage of some new noble blood.

After a few days, O'Shay's body begins to heal from the toxin of the spider. He is lucky, the doctors tell him, that he noticed the bite of the notoriously shy spider and got to the hospital before deep rot had set in.

When he interrogates Gordo, O'Shay is struck by his calm as he faces the loss of all his cars and money, by his civility as they talk. They are like brothers staring into each other's dead eyes.

HE IS LOOKING FOR ROOTS and so he comes to the Civil War burying ground. He stands over his ancestor's grave, the ground green and seemingly at peace. A creek runs nearby. And then he feels the closeness that he can't quite describe even to himself, the presence.

That smell, he glances up from the grave, yes, that stench of the creature. He wonders if the thing haunts him. Or is itself haunted. He knows this: the thing does not have malice, but it is malice. It comes from a realm that Joey O'Shay does not want to think about.

He knows that it is here at the grave, and not because of some long-forgotten battle, or because his ancestor owned slaves. He can smell the emptiness and loneliness. He can feel the weight of that black cloud when he was a child dying in the back room of the family home.

He does not know what it is.

But he knows that it is real.

He wonders if it follows him alone.

But he knows it does follow him. He cannot deny this presence on his trail.

O'SHAY WANTS THE AUDIENCE to understand, or at least that is what he tells himself. He gives them glimpses of himself as a boy so that they will see where the man came from.

He saunters down the hot streets of the city to a little local store. He buys a cola. The man behind the counter also sells heroin. That's childhood. Some of his running buddies grow up to be dealers. Some become solid citizens. Others become convicts. Some settle down as addicts.

O'Shay takes his own path. He starts drinking in the eighth grade. He lifts weights. He smells the honeysuckle as he sits with his girl on her porch. He floats in the jon boat, sees stars overhead. He drinks in nature in the summer, runs the streets in the winter. He is like the children in *To Kill a Mockingbird,* who in a world of evil, cluttered with dangerous adults, face down evil with honesty.

The audience sits silent and puzzled in the big room of Bobbie's hotel.

O'Shay leaves childhood, races on through the stab at college, the marriage, the stint in the service.

He moves furniture by day, hauls silage at night. His wife is with child. He is standing in his mother's yard when a squad car pulls up. Shit, O'Shay thinks, he is used to being hassled by the fucking cops.

The cop gets out, and O'Shay knows the face and voice, a kid from the wolf pack he used to run with.

The guy says they are hiring. O'Shay thinks, I got a woman and a child, I'm making a fucking buck and a quarter an hour, I'm working two jobs and I'm on the edge of doing something that is not police work.

O'Shay looks up at the audience and says, "I had no intention of being a police officer."

The man he hauls silage for, an old guy with a dairy farm, takes O'Shay aside and says, "Don't let it break your heart or make you mean."

O'SHAY KEEPS THE VIDEOTAPE of Popsicle's torture and smothering in the back room by the French doors that lead to the patio where Joe the Crow visits, where Geezo bangs on the glass demanding food. At night O'Shay sits on the black couch there, sipping a drink, the only light a glow from the other room. On the walls, paintings of beaches and waves, that water thing that seems to calm him. The house is a museum of paintings of the sea.

On rare occasions, O'Shay will show someone the tape, although he's leery of people getting the wrong idea from the video. Every time he pops it in, he moves a chair so that he is sitting right by the screen facing out. He can see nothing.

O'Shay has watched the tape only once.

He turns the killing over and over in his mind looking for some tiny crack in the event through which he can slip so that he can put his large hands on the throat of Q.

Another item on his to-do list.

He wants to see Q on a gurney as the needle pumps poison into his body.

OUTSIDE THE SKY IS GRAY and this is good. The city thirsts for rain. Inside O'Shay stands by his lectern facing forty-eight chairs, row after row of contented faces. The chairs face tables with nice green cloths, with pitchers of ice water and glasses. Bobbie has seen to everything. If O'Shay wants to do this, okay, she will provide the setting. But she refuses to listen.

She will not tolerate this talk of getting out.

She knows if you are truly in, there is never an out.

You can only leave if you have never been there.

You will miss the action, the rush, for one thing. You will not know what to do with yourself. But there is another matter, one the fucking counselors and therapy scum will never know.

You cannot leave because you have learned things that chain you, that own you. You have tasted death, guilt, and, more deeply, you have tasted violence and this taste does not make you recoil, does not give you nightmares. This taste floods your body and becomes appetite itself. You cannot stop seeing once you know, you cannot stop feeling once you know, you can never stop the pain once you know. And nothing, no amount of drugs or booze, can erase this knowing.

You cannot go back, period.

And Bobbie thinks no sane person would wish to go back.

No one, in her eyes, would willingly give up the knowing.

So she stays in her office, works the phones, steps outside now and then for a cigarette. But never, for even a single second, looks into the room where Joey O'Shay plays out his life like a spool of a film.

He is up in front of the room. He moves heavily, his body is mass, solid, and his tread slow and sure, yet light. He is a huge piece of furniture moving in the front of the room but he makes hardly a sound and has a rough kind of grace.

He is telling the audience the story of the professor, the old wino, who translated for the deaf-mute Indian from Mexico, stunning the court. He explains how drunks are just people who have been broken by life, how they are merely a few steps and moments removed from anyone in this room.

Years later, he continues, he picked up the old wino professor again. He searched his pockets and found in them a dead bird. And a razor blade.

O'Shay looks down at his hand and sees the tiny scar he still carries from nicking that razor.

"I'VE NEVER NOT BEEN EXPLOSIVE. I was just a no-good motherfucker." O'Shay sits outside at a café on a blues street where the white people come to feel black. When he was coming up this was the district of low-life bars. Now it is where the children of the privileged come to taste lives they are afraid to live. He licks his lips as he remembers the slippage, the savage beatings he dished out, the drug buys without backup, the craziness. He sips coffee, fields calls from informants, tries to get to the place where he can make sense of himself.

"There is a point," he begins, "when you are on the street so long and you learn so much ..." and then he stops, his small eyes get smaller. His jaw tightens. He starts again, shifts into new territory, "You go to court, you see some black kid, who looks like a little kid, get ten years for three or four rocks of crack and the next case is this old white dude who goes up to the bench with two high-dollar lawyers and the old dude's grifted more than a million from some fucking savings and loan and he gets a fine and eighteen fucking months and I look over at these black people, probably all relatives of the kid, and they roll their eyes at each other, and my partner leans over and whispers to me, 'What the fuck. He grifts more than a million and he gets a fucking fine and a little jail time.'"

Then he stops again, sips his coffee, and gives up on explaining.

THANK GOD, SHE HAS FOUND a way to use bright colors. She craves them after all the gray and muted canvases. This time Cosima clearly delineates the background strips of film with sprocket holes in each strip. The film is blue and green and yellow and red and lavender. The film looks like a berserk rainbow that consumes the heavens. In the foreground is a clapboard with a box for the production, the director, the scene, the take, the roll, the date, space is given for all the details.

The clapboard is open, the top bar lifted and about to bang down to start the scene. It looks like an open jaw ready to devour. And a yellow band of film streams down into the clapboard, and on this film are scenes, the smug bearded man is there, a plane approaches the dome of the Capitol, towers burn, crosses rise over graves, a world dies. And yet below the clapboard the filmstrip does not continue. Apparently this is a movie consumed in the act of production, a one-take film that ends in ash and death.

Cosima stands back from the canvas. She has created her ultimate dream: a deal, a moment, a finale where everyone goes down and nothing, nothing at all exists after the shoot is completed.

THEY MOVE THE JAIL, toss up new cells, and work out a design that will make everyone feel better as they do their time. O'Shay sometimes visits the old jail, now an empty ruin. He stands in front of the admitting desk, shuffles his feet, feels how the old wood is worn down from the thousands who have waited here to become numbers.

He is convinced that the old jail is still full.

He says, "You can hear movement up on F block."

He refuses to describe the sound and falls back into himself.

He cannot seem to tolerate going to church. He tries but there is nothing there for him. He'll hit a mass, treat it like a wine tasting, but after the mass he is still thirsty, still seeks stronger drink. Life means something, all these things he has seen and endured mean something, the lost souls are still moving on F block, but nothing he hears says what he feels the meaning is.

O'Shay has a habit. He will start a story, then slowly sink into its web, his voice will flatten, always it flattens when he reaches the zone, he squeezes his voice until it is incapable of excitement, and he slips from the past tense to the present, he relives what he is saying.

And then, every time, he breaks off his story and says the same thing.

"Don't mean nothing."

VIKTOR FRANKL IS DEAD NOW but not to O'Shay. The wife the man wondered about, she died in the camps. He remarries. He writes the lost book, only now it is a different book because he has been there, been to that place O'Shay knows, that place O'Shay insists he is finally leaving.

The man is in the sixth winter of his war when F comes to him. Frankl notes: "I once had a dramatic demonstration of the close link between the loss of faith in the future and this danger-ous giving up."

F had been a composer. Now at the beginning of March, he tells of a dream he had in February and in this dream the camp is liberated on March 30th and the suffering comes to an end.

F believes his dream, he is bursting with hope.

In the camps the days are very, very long and yet the weeks seem to pass in seconds. Everyone in the camps remarks on this new and curious sense of time that pain brings to them. The news that dribbles in from beyond the wire is bad, very bad. The hope of liberation fades, the reign of pain seems endless and secure.

On the twenty-ninth of March, F suddenly falls ill, his fever soars. On the thirtieth he becomes delirious and loses consciousness. On the thirty-first he dies. They say typhus killed him.

On a break, O'Shay goes outside the hotel. He is fidgety. His performance has been flawless, not a single stumble, not a single note in his hand, a pure flow from the reservoir of his life.

He looks up and says, "You don't think I know I am dying. You don't think I know I can't go on living this way. If I don't change, I'll be dead in a year. I know that."

O'SHAY JOINS DEEP NIGHT, the shift when only people like himself are out or people like himself who do not wear badges. He looks at the audience and suddenly he is like a recording with a skip and in almost the blink of an eye he has recounted or alluded to his first killing.

"There are people," he says very softly, "who will kill you for no reason but simply because you are in their way. My whole life changed. I joined the realm of people who had done that to someone."

He begins to riff, the shootings and killings lose their moorings, float as pure and distant things in some menacing sky that only O'Shay can clearly see.

"As she fought for her life," he intones, "she smelled fear in her nose."

O'Shay is in a bathhouse and fifteen men are having sex together. "I was terrified."

The marijuana war, nine dead in its wake.

"A man," he says, "lying dead with his eyes open and a .410 shotgun in his hand. His friends are shrieking and screaming."

He drags a fellow officer to safety, looks down and sees the man's intestines spilling out. Another officer and the man begs, Look, look, did they shoot my dick off?

Angel Sonrisa is back too, and Joey O'Shay tells the audience what Sonrisa taught him. He told him, "Be yourself or you will go mad. Always use your own personality or you will destroy yourself. If you try to be a badass, a bully, someone will kill you."

He is back in the night, the deep night and Sonrisa takes nine rounds, "Oh, shit," and he dies and "a punk who looks like a baby killed him, a thin little weasel."

O'Shay pauses, he does not stumble but he pauses, he sucks in air, he is close now and he looks out at the room of passive faces and pension eyes.

"The guy," he says very slowly, "the guy who taught me the most was killed and he became one of a series of people around me that became dead or paralyzed or out of their minds in an undeclared war."

He tries to explain his life after that, being the entry man, first through the door, gun in hand.

He approaches a door, a woman leans out her window across the street and screams, "Police a-comin'," O'Shay plows forward, a man leans out the window of the house and pumps rounds into the gut and head of the cop behind him.

"I remember seeing the back of Jack's head was shot off."

The voice is so soft now, the voice of a lover.

O'Shay looks at the massive blood flow, reaches down and tries to put the back of Jack's head on.

He gets up, and goes forward. A man throws his artificial leg at him, in the back of the house an old woman reaches for a gun. A cop gets a thousand stitches in order to hold his guts in. The people say they thought the police were thieves coming for their property.

The audience cannot keep up with O'Shay now. And if by

chance they could keep up, they would refuse. This is not their country.

He tells them that if you hesitate, you die.

"I had a son die," he says out of the blue. "You will do anything to escape that anguish. I sat and watched my little boy die and I had to find some escape and kicking down doors was not enough. I didn't give a shit if I lived or died. Two years living in a shithole apartment, buying from more than two hundred people, no backup, no surveillance. People shooting up in front of me."

He stops. He looks at the floor. He looks up.

"I had in my mind that the moment someone tried to make me use drugs, I was going to kill them.

"LIFE WAS CHEAP," he says.

He rooms with one of the biggest meth cookers in the region.

"I deceived him. It is a sickening feeling to deceive these people. I felt ashamed. You become so close to your enemy that you become him. The loneliness I felt was killing me."

He is crawling over a fence, moving through the night, he is light on his feet. He reaches up and puts presents in the knothole of the tree for his boys. And then vanishes. It seems like a movie seen by a child.

His partner breaks under the work.

"When I got him, he looked like a little kid. When I finished with him, he was jaded, almost alcoholic, and burned out. Two years."

He looks briefly at his audience. He knows now he is alone in this room full of people.

"I was mean as shit by then, I was mean and reclusive."

He tells of his discovery, his invention, the unknowing informant: "The most ugly philosophy I know."

His face now is blank, a wall made of flesh and no chisel on earth could carve an emotion on it. His voice is still low and soft, but everything is rushed now, flash cards he flings in the face of the audience. He cannot linger now that he knows he is alone in the room full of people. That there is no one to hear him out but himself.

The girl, the loser, so young, he says, "Hung herself in a fucking hotel room."

HE GETS UP FROM HIS office chair, goes to a file cabinet, hauls out a small green box, and opens it while the still air fills with the hum of the fluorescent lights. The carpet is gray, the desks are spotless. The Medal of Valor hangs from a blue ribbon. The back of it is stamped STERLING.

He thinks of that moment: dragged two wounded cops out of the range of . . . And then he falters, and does not let himself go on.

He never wears the medal. The surface is tarnished with neglect.

SUICIDE IS FORBIDDEN. It violates the order of the camp. But it is also forbidden to save a man who attempts suicide. He is to be left dangling from his rope.

So it becomes important to sense when a man is on that edge and to do something to pull him back to life. Frankl learns to find a future for such men. One man has a son in a foreign country and so he fills that man with the sense of his son. Another man has written many scientific books but his work is not yet finished. This is pointed out to him, that only he is capable of completing this necessary work. He, too, decides to continue living.

There must be this belief in a future.

Joey O'Shay thinks this is easier said than done.

THERE SHE IS UP ON THE SCREEN, that girlish smile, that toss of the hair, the whole scene glows simply because she has walked into the lens of the hidden camera. The audience peers at the ill-lit, grainy tape, tries to catch the poorly recorded dialogue. O'Shay stands off to one side, his face a blank. He commands attention and yet is never animated. He violates every rule of public speaking save one: he has something to say.

Gloria beams at the placid civil-service faces in the hotel meeting room. She is in her glory, the smart woman from New York meeting major drug connections in the city. There is Joey O'Shay greeting her, and now she sits on the sofa facing him.

O'Shay tells the audience, "This lady is very sophisticated, very nice, very polite."

But he says no more.

There are things he will not tell this room.

Earlier he'd told them, "They're evolving like a virus and we don't have the antibiotic."

He seemed to be talking about people in the drug world. But he never clarified this point.

VIKTOR FRANKL THINKS HE comes back. He believes that there is a moment when he shed his death and returned to his life. But then Frankl insists that a person can turn any hell into a ticket back to the safe and green places. That is his hold on Joey O'Shay.

The camp has been liberated, the meadows are full of flowers, and the man walks mile after mile into the first real spring he has tasted in years. Birds, he hears birdsong, larks lifting toward the sky. He can taste space, he can walk any direction he chooses.

The man falls to his knees. He sees one sentence, this string of words suddenly filling his mind: "I called to the Lord from my narrow prison and He answered me in the freedom of space."

He keeps repeating this sentence over and over and he loses all sense of time as he kneels on the fresh earth of his new Eden.

Later he decides that is when his life restarted and he began his long walk back to the place where he was once again a human being.

Joey O'Shay envies the man. He must find his place to kneel, his ears are eager for birdsong.

LUNCH IS AT A MEXICAN JOINT in the area where he began as a beat cop so long ago. He remembers a night back then, when he'd been cruising an alley in the late hours, and found this guy in his car, some kind of car trouble.

He tells the guy, "Look, this isn't safe here. You're in back of a queer bar and some mean fucking faggots hang around here."

The guy laughs and says, "I own the bar, I'm a faggot."

O'Shay falls all over himself apologizing.

The man laughs again, his brown face happy with himself.

That was a long time gone, back when O'Shay could cling to us and them, to that sense of otherness.

Now that is gone and he knows it.

After lunch, the rain falls like a fury, the hungry ground drinks. The creeks are rising, he can smell it.

HE GOES BACK TO THE HOTEL, but he never really comes back. He stands there and talks, he shows surveillance videos—O'Shay sits in a chair, his expression blank, as a man goes down on the floor, the machine gun pointed at his back, terror playing across his face as he thinks he is about to die.

O'Shay says as the tape rolls, "That's how fast their lives end for them."

His voice is as flat as usual but there is a hint of sadness in it.

He's found something missing in the room, something wanting, and so he refuses to give the room any more of what he is.

When the afternoon session is over, he drops into Bobbie's
office.

She looks up and smiles and asks, "Well, how did your little
thing go?"

O'Shay is ill at ease, he is not in control at this moment.

He says he guesses it went okay.

She smiles at him again.

Her eyes say, You can't get out, you can't get back, you can't
ever be a fool again. You must play it out, there is no choice. This
is the law. Once you know.

The creeks, they are rising, rising all over the city as the gray
afternoon becomes the night.

O'SHAY CRUISES THE STREETS where he was raised, streets of shotgun houses shaded by big trees. Each creek reminds him of his
childhood fishing days. He stops the car, looks down at the
wooded and meandering stream, and the smells of his childhood rush into his mind. He can see a child down there and the
child is himself. But he does not get out and go down there. He
can't seem to do it, he has never returned to the creek except
after midnight in his mind.

The faces are all black and brown, much as they were when
he was a boy. O'Shay drives around for hours, he revisits old
football games, the church where he and his friends boosted
chalices at age eight and then decided they'd all go to hell so they
returned the loot. Ah, there is the field where he got tangled in
the barbed-wire fence when he was running from the cops and
took a hard blow from a nightstick in the gut. There is the gas
station where a cop got pissed at him and made him lie on the
ground on a freezing night while he was sprayed with water and
left to shake with cold.

He has returned here in part so that he will not feel alone, but the old neighborhood only reinforces his sense of aloneness. This sensation is hard for O'Shay to talk about or examine. He has spent his working life as part of a team, a unit, a bunch of guys, and he has made friends and he has admired people and he has been taught by others. But he remains alone and he knows it. He drinks with people, with friends, but somehow never lets them inside of his mind and his real life.

There is that day over lunch when Bobbie is talking about O'Shay and how she knows he's crossed the line, how she knows he's evil, and she tells how losing a child changed her life and how she determined never to let anyone get that close to her again or hurt her that bad again. And when it is suggested to her that losing a son hurt him that way, too, she says, "Son? What son?"

He works with people but they do not know of Joe the Crow or Geezo or of his painting, much less of that boy whose name he can barely even whisper to himself. It will be a Friday night and he'll go to some bar for another cop's bachelor party and he'll drink and laugh and talk and fit right in. But still be alone. He cannot find a way out of this aloneness.

That night when O'Shay was doing his big heroin deal and the phone rang for hours, always a hang up, one of his grown sons was over. O'Shay got out his guns because he thought, Maybe they are on to me, maybe this time it is all coming home. And his son left in disgust because Joey O'Shay had brought it home, he could not ever seem to give it up. The father who left presents in the dark of night in the knothole of a tree was suddenly looming before him again and he did not like the memory.

That's the problem with the word *alone* and the word *lonely* and the word *loneliness*. O'Shay has entered a country where the rules are that you must be alone.

He remembers when his boy died, and then the bad times came and the viciousness poured out of him like lava, and the killings followed and the beatings and the sense of being numb to what he was doing but not near numb enough to what he had endured, he remembers that time and how, finally, he realized he was crossing lines, lines inside himself.

So he holed up for three days and wrote, tried to sort it out, to make some sense of the evil he felt within himself and the evil he saw around him. And the dead boy also, he tried to face that.

Maybe that is alone. Or maybe that is finally facing the awful fullness of life.

O'Shay wonders if the old snapping turtle he saw at dusk as a boy could still be alive. It's possible, he thinks, and if so, it's still hunting the waters and no one probably knows it exists. Except Joey O'Shay.

HE READS ABOUT IT in the newspaper, not on the front page, but tucked away in the back. It is not an important story but O'Shay clips it and keeps it because, well, he's not sure why but he is certain that it matters to him.

Someone has hauled out a monster catfish, a beast weighing in at 128 pounds. Caught the thing with simple bait while fishing from the bank of the lake right near where Joey O'Shay used to put out his trotlines, where he would float in a jon boat at night and take in the stars. The fish in the story is five foot eight inches long with a twenty-nine-inch waist. The guy who catches it is twenty-seven and the fish is probably the same age as well as the same size.

The fisherman realized what he had hooked. He jumped into the lake and rolled the fish toward the shore. He called a friend, said, Hey, bring a hundred-pound scale. He huddled in the water, holding the fish for an hour as he waited. He was determined that anything that had lived that long and grown that big would not die at his hand. But the scale proved too small. So then he put the cat in a box and hauled it twenty minutes to a big scale.

The cat somehow stayed alive. Now it swims in a state fishery exhibit, tamed but alive.

Been floating under O'Shay's boat all these years in its own universe, growing and feeding out of sight while the world had its wars and laughs and pain. Until someone trailed the bait before its face. And now saved by a fellow predator who could not bear to slaughter such a big catch.

O'Shay puts down the paper. He has read of a clean thing.

JOHNNY BOY HAS GOTTEN five state and eight federal and the terms run one after another and this means at least ten in the joint for him no matter what good-time bullshit the system offers up. O'Shay figures that long before then, he'll nail him for some murders and keep him there until he is too old to return to the streets and kill some more. He's got time, he'll seal Johnny Boy's fate in due course.

That's not what is bothering him. It's Gordo, Johnny Boy's uncle, the drug merchant who lived like a prince in that block-long compound. He remembers the night he interrogated Gordo, and the calm and dignity of the man who was facing forever in a cage. He remembers almost feeling like a brother to Gordo as he took his fate without complaint or whining.

Now Gordo is making noises about wanting a trial. This is an affront. He is nailed, taken down fair and square, all the letters dotted and crossed. He is dead meat.

O'Shay and his crew are revolted by this whining attitude. He will cane the son of a bitch, he will flog him, he will rain charges down on him like a murderous hail. Trial? What is this shit?

There are rules and when you are cleanly caught and hung out to dry, you are not to complain. O'Shay prides himself on building complete cases, structures whose roofs never leak, whose doors slam shut.

It is a matter of morality for him.

When he was closing out all hope for Gloria and the others caught in his snare, all those Dominicans and Colombians and Mexicans, he had to meet with two bosses from a federal agency.

O'Shay goes. It is Halloween, and one is some kind of troll and the other a candy-ass vampire or something. One is a man, the other a woman, but to O'Shay both are ridiculous. There is a decency on the street, some kind of honest ground, people are ruined, Gloria is sitting in a cold New York cell, O'Shay has put his life on the line, and here he's talking about what to do and how to settle the case with a troll and a vampire. And he and his crew feel violated. He feels sin for being there. There are decencies that must be observed, things old cops taught, and criminals and creeks and stars at night, and you don't sit in an office dressed like pansies and decide the fate of hardworking drug dealers. It is a violation, an insult to the crew.

Music out on the street, hard songs, soft riffs, killing percussion strokes, dreams floating lazy in the curls off cigarettes at two a.m. in the suicide bars. There is a world that is rock hard and actual and it runs like a stratum through the city. And there is this other world, soft and dishonest and with tricky accounting systems and thieves with white collars and banks in their pockets, and it is worshipped. Johnny Boy's going away, poor bastard doomed almost in his cradle. Gordo, the riches over, the fancy cars in the garage seized by cops so they can tool around and play at being bad guys.

BOBBIE KEEPS ON LIVING. The hotel management grows and she prospers. She dabbles in her secret life, a charity run by her that gives toys and outings to kids in halfway houses, kids with troubles but without parents. She holds banquets, shakes down businesses, she is the Santa Claus of this world. And yet apart from it, a force that produces results but remains largely hidden from view. This suits her feel for things.

Her health problems become a private matter because she is inept at being a victim in any circumstance. Her smile never stops, not when Gloria goes down, not when O'Shay closes out his case and fills cells with his heroin dealers. Not even when O'Shay once again tells her he is hanging it up. She smiles at all of this because she loves the hunt and because she does not believe O'Shay can quit. Because she cannot quit, cannot quit her habits, cannot quit her body, cannot quit the way she prefers to live.

And if she cannot quit, he cannot quit, not in her mind. Once you commit to a life, her eyes say, you must play it out.

HE GOES DOWN INTO the storm drain, the air dank, his nose full of stench, and everywhere darkness. Ahead, he somehow picks out two eyes, panic fills him and he runs.

He fetches his mother. She goes into the drain, hears a noise, and calls the fire department.

They come out with a four-foot alligator.

O'Shay looks at his audience in Bobbie's hotel. The story lies there, unexplained and yet somehow thrashing and menacing the darkness of his childhood memory.

He tells them he walked a mile to the barbershop at age four and the world was not dangerous.

He is moving now, slowly prowling from one side of the room to the other, that catlike grace belying his solid body. He's back where he began, back at the creek, the swamps, the sewers, back when the world was safe, and yet there were these things, largely unseen by the workaday world but still alive and actual, and these things have traveled with him through everything else that followed. That's the rub. The world seems safe and sound and yet this very hotel for years has been Joey O'Shay's den of drug deals and drug dealers and machine guns and cameras secretly watching and white powder on the table and piles of fresh, green bills.

THE ROOMS ARE SAFE and also perfect. Bookshelves of fine wood, rare porcelain figures of vanished empires in glass displays. Paintings smile from the walls, and the furniture hardly makes a dent in the large New York rooms. The women dress well and are graced by large and bold hats. The suits of the men never wrinkle and everyone smokes in a most elegant manner. The bathtub is a marble vat, and there is fine liquor for everyone all the time.

Laura herself glows with innocence and beauty and wealth and ease, her face so smooth it seems lifted from the blank gaze of the porcelain figurines. The detective has suffered from machine-gun fire but has come through this agony and now has not a hair out of place. He dresses as well as a corporate chief and says little, in a flat voice. He is clearly a man who knows how bad life can be and wishes it were not so.

The plot is very smart and pointless, a dance around guilt and, of course, Laura at one point is implicated. But this cannot be, any more than the detective can help falling in love with her. First, he falls in love with her face and memory when she is believed dead—mutilated beyond recognition by a shotgun blast to the head. Then, when she turns up alive, he falls in love with her manner, her scent, her kindness, her lack of guile, and, of course, her innocence, somehow maintained even as she maneuvers in a depraved world.

Joey O'Shay watches this film endlessly. He knows the cop is a fiction. He knows the rooms and fine things by rumor. He knows the plot is a pastiche of false suspense. He knows the high society of the movie is as distant to his own life as the court of Louis XIV. But he knows the feeling the movie imparts, the feeling of innocence being menaced by evil. And the film is black and white with little space for gray. And the cop in the film does one thing that catches the eye of Joey O'Shay: he never gives up, regardless of what pain lurks in the answer he is seeking. That tells O'Shay the cop knows what knowing means.

Most of all, Joey O'Shay knows what he needs and he needs *Laura* just as he needs the children in *To Kill a Mockingbird* finding presents in the knothole of a tree.

HE FINISHES OFF THE HEROIN CASE, more people fall in New York, Colombia, Mexico, a clatter of broken lives shattering when Joey O'Shay pulls the trapdoor under their dreams.

Life is flat. He hates his work. He cannot speak to his colleagues. He says he is finished. He finds it difficult to move.

But he is ready. He can read a face in a glance, know a move before it happens. Sense what someone else will do before the thought crosses their mind. His reflexes are sound, the tools sharp and pointed.

He has this thing he is good at, and it has no listing in the phone book, no tiny-type notice in the want ads. But it is needed, almost begged for in the place throbbing around him and ignored as the cars speed past, windows up, doors locked, arms at ready.

He can protect the fools and the liars and the rich and the gutless and the straight with their crooked ways. And that is all he can see and all he knows how to do to his satisfaction. It is not about money. He has more money than he will ever need. It is not about a job. He loathes his job. It is about something the world has grown deaf to, a calling. And the calling is hard and the calling is not pretty and most people, almost everyone in fact, want to know nothing about the calling. Until they need him and his skill.

A spate of bad killings begins, reaches out past the city, bodies left all over the region, someone going around with a list and snuffing out lives one by one. They ask O'Shay to look into it.

He stirs. He nails the first killer in five days flat.

He was still a beat guy in uniform, his one-year-old son barely dead, when he had his first deadly confrontation. There was a brief flutter in the department about making sure their man was okay. He almost died himself at a young age when he was delirious with fever in the back room of the family home and sweltered through the night with the Evil Creature hovering above him and pressing down. Like his own parents, he was helpless to help his own young. He can make all the obvious connections except to believe the obvious connections really explain anything.

He believes in spring, like Viktor Frankl in the book, he believes that he will come down the path, buckle under the fresh weight of flowers and birdsong, and begin his return to some better self. He will shuffle off years of fists and guns and betrayals, shed all this like some snake and then move on with a new sensitive skin and the sun will bounce off him and warm his heart. He will start painting again, the workouts will flush his body of poisons, he will plant okra and peppers and tomatoes in the garden and return to good food and decent living. He will cut back on the drinking, and the rage.

He will go to the lake, back to the calm waters and the stars, back to where he first got an inkling of what life might be and should be, to a place of color and form and scent and sudden stirrings in the water as the big beasts move on their rounds, innocent and ignorant and yet exultant. And he will not die.

Yes, he will go to the lake and sort it out. He will not take a vacation, he hardly knows what a vacation is. But he knows comfort, the feeling that comes when you enter ground that speaks to you and you know instantly what is being said. The lake, the cove, the boat. The stars.

He knows he is lucky. The answers are silent, the mysteries stream by his blank face, the body, coiled but eager, hurls forward.

It all means something or it will mean nothing.

He says, "I've been a cop thirty years and three weeks. I had no intention of being a cop."

Then he pauses.

He looks out at the audience and he loves them in some part of himself, he loves their faces, he loves their homes, he loves the lawns they tend, he loves their children, he loves their safe and sure ways. His eyes may be blank but, back beyond this shield he has constructed, backstage from what the audience can see, his eyes flood with his heart. He loves the certainty of the faces before him and envies the lives they return to when they leave this room. And at the same time, he cannot desire to be them. He is something else, something made in part out of his own being and in part out of what he has tasted and had to digest.

Finally, he whispers the last thought, "I got sick of it. I have to step back in time and go over my life."

Stars

12

BLACKJACK OAK AND CEDAR ramble the shore of the cove. Night will come and then the stars. But now the day wanders the still waters and silence hangs like a sheet over everything. Pelicans gently rock, gulls gorge. Balance has been lost, a golden alga bloom radiates out and half a million fish bob dead on the surface.

Before one of the wars they planned the dam, and then during that war they used prisoners to clear the ground. For years after closing the sluices, the waters slowly rose, and now the lake is a snake with dozens of fingers probing into the land and creating coves. Small islands beckon and the sky above flutters from blue to scud clouds to moments of gray and light rain and then back to blue. Nothing is stable at the moment except the vultures roosting in large trees on a cliff. The forest looks largely barren, the limbs stark but with buds swelling and whispering the promise of spring. Green grass creeps among the thickets and raccoon tracks edge the waters.

Joey O'Shay has come home.

He is back where he began, the cove called Paradise on the waters trapped from a murdered river. His eyes squint as he takes in his place. His face remains a blank. He visits now and then. But he never leaves here.

Foxes scramble through the brush, deer browse, and a rare mountain lion moves quietly through the woods. Ducks come by the thousands, geese also, and for moments or hours the world can seem to have order and order can seem to be natural.

There came a summer when Joey O'Shay was eleven and his folks dropped him off here at the end of school and came back and fetched him in late summer when the city once again shackled its young. For months he lived on his own, a rifle, a trotline for catfish, the smell of the woods after a rain, stars working the night overhead. He and his father had built a windowless storage shed, maybe ten by eight, a cell in the woods, the walls painted a pale green. When storms swept through, the boy slept in the shed. The rest of the time, it was sky and sun and stars.

O'Shay still keeps a photograph of himself at age four, a small body with a large head, the eyes staring at the camera with seeming incomprehension. Before him is a birthday cake. The child in the photograph does not smile, the eyes look cautious and yet surprised. O'Shay has kept the photo close at hand for almost his entire life. When he looks at it, he catches that first evidence of his state of mind: he is an alien and his family and those around him make no sense. In the photo, he sees his life: alone. Not isolated, not shy, not alarmed. Alone. He can listen and understand words, he can make the moves, catch the ball, run through the green grass of summer. He learns readily, succeeds at tasks. But he lives alone in a space behind those eyes, a space where all the scents and colors and sounds of life constantly pour in, a space where he can hear what is really being said, the words behind the words, he can see what is really happening, the slight unconscious signals given off by others without their knowledge. He can look into someone's eyes and know what they are thinking as their mouths lather him with lies. He is that child, older now, but still that child. Now he hunts people with all the faculties he first became aware of as a child. And now as then, he is alone no matter how many people surround him.

Last night there was a party in the city, the bar teeming with cops and lawyers and other feeders off the system called justice. O'Shay brought the party into being, rented the bar, arranged for the band, beckoned his world into some celebration of what they are and what they do. The party was ash in his mouth and yet as he walks the woods now, he knows the party was essential as was the ash.

Crows call out, a fish jumps, the sun burns and then flees behind low clouds.

The party lingers in the air, even here at the lake, lingers like the haze of cigarette smoke in a room that has emptied of people. The party is a last hurrah, though the occasion is unannounced.

Bobbie beams at the granite bar, drinks rum and cola, lights a cigarette beneath a no smoking sign, and offers, "He despises every person here."

Then that smile of quiet satisfaction. Her mind glows with the energy of cocaine as swarms of narcs cluster around her. She is here at O'Shay's invitation, at his insistence, for reasons never stated but understood. She is to be the witness to the real meaning of the party. The four-year-old staring out over the birthday cake, the child who comprehends that others do not comprehend, that child is now a man and that man wants someone here who realizes how alone he is and what he must deal with. And at the same time, that man needs the people in the bar, needs the sense of a blue brotherhood of cops, needs them even though he barely belongs to this world to which he has given his life.

Bobbie surveys the bar and sips her drink and says, "I'll tell you exactly how he will be killed. I've met the guy who is going to do it."

She says his name, this killer's name, tells of the coldness in his eyes, the way he moves with catlike grace, the way he paces alone in his room, paces day and night and never leaves his room and never calls anyone but Cosima. He has been smuggled in from Colombia and everything is very secret. His phone is tapped and of course cameras watch from tiny pinholes in the walls and the clock by the bed and from bases of lamps and from the ceiling, cameras watch and watch and the man does nothing but pace. No words, no expression.

Bobbie says again, "I tell you he is the one. And she is the only one he calls. And now she insists the man will only talk to Joey. She is orchestrating it. This guy is the one, this guy will finally kill him."

She beckons to O'Shay who is glad-handing his way through the crowd. He's been drinking for hours but he seems just like he did when he first came through the door.

"I've met him," she tells O'Shay, "met the guy who will kill you."

His face becomes even more blank.

"He's the one," she insists, "he's going to do it."

O'Shay says nothing.

"I mean it," she adds.

"He is a killer," O'Shay allows, his voice flat.

"Well, he's the one."

"I've had a good life. When it's your time, it's your time."

His face reveals nothing, the muscles relaxed, cheeks smooth, brow polished like marble, eyes open and yet slammed shut to any questions. He is not cutting her off, he is not pretending disinterest. He is simply not reacting.

Bobbie exhales, smiles, and says, "I've decided to have a drink in your honor three weeks after you are killed."

He says nothing, not a word.

He offers no explanation of the Colombian, not why he was secretly taken out of his own country, not why he was secretly brought to this city, not who or what he really is. He stays blank and yet Bobbie can tell he is not hiding something from her or himself. He simply does not care.

O'Shay holds a longneck in his hand, his eyes squint, his face remains immobile.

Bobbie simply smiles back.

"You wait. You'll see I'm right. He's the one."

The party swirls around both of them.

O'Shay has cut Cosima loose, he has stopped answering her calls. She is on her own. She has not been invited to his party.

He sees two or three futures for Cosima. She will get caught doing a drug deal. She will be murdered by someone, one of the several men out in the night of this city who has accepted a contract to kill her. Or she will be deported. No matter. She has ceased to exist to O'Shay and once that happens, no one is ever recalled into his life. The child in the photograph, that four-year-old, that person who knows he is alone on this planet, that person goes through doors easily and he never comes back.

He allows that Cosima could set him up to be murdered. But she will go, he has decided. Just as he will go. After the Colombians, after the massive heroin deal, after Gloria, O'Shay has determined he is through with the way he has worked and lived. And now he is hosting a party to celebrate this decision, though no one at this party, except Bobbie, knows the reason for the event.

Irma is doing time now, her perfect body with tattoos of green and pink to accent her crotch wasting away without a man's hand to caress it. Garcia is caged in Colombia and O'Shay does not even have to ask how he is being questioned. Gloria is in a prison, doing her years, perhaps now and then when time gets heavy, remembering her tender moments with her soul mate, the name she still refuses to give up. O'Shay is leaving all that behind.

He grabs a large black woman and begins dancing.

Bobbie returns to her drink. She has said what she wanted to say. Her face glows with pleasure as she leans against the bar, sips her rum and cola, jokes with the narcs around her. And makes frequent trips to the ladies' room.

She floats in memories, and they come racing like horses, fly out of her mouth in the din of the bar.

O'Shay is a young buck when he comes to one of her properties and rents an apartment. He is bearded, wild, and people go in and out of that apartment all day and all night, buying this and that, selling this and that, sealing their fates with a laugh, a beer, a shot in the vein.

Two doors down from O'Shay's place a black pimp and drug dealer operates, runs his stable of whores out of his apartment and moves his dope. One day, Bobbie finds the unit surrounded by cops, SWAT teams with guns out, all focusing on the pimp's apartment. He's shot a cop and then fled to his place. The cops tell him through a bullhorn to send out a woman, some fourteen-year-old he is running on the street. And eventually the girl comes out and they take him down. But that is not what sticks in Bobbie's mind. It is this: all through this takedown O'Shay is two doors away running his scam in his apartment and not a damn cop on the scene ever makes him, or his enterprise. And not a damn cop knows he is another cop running an operation under deep cover.

That's the O'Shay she knows, the guy who moves through her city unnoticed. Except, once in a great while, by some of the people he destroys. That is the guy she watches as he dances now in the bar.

The band plays,

Give me the beat, boys

She knows who is going to kill him, and it somehow comforts her at the same time as it disturbs her. What disturbs her is that a world without O'Shay will seem empty. What comforts her is that a world without O'Shay will be more her size.

She is tired, it's been a long day, but she knows she cannot leave the party because she cannot leave him here all alone.

One time a partner of O'Shay's came to Bobbie and said, "If I stay with him, I know I'm going to die." And so he left, fled back to some part of the police world that seemed more sane.

Now another former partner comes over to her in the bar, a short muscular guy with huge tattoos flowing up his arms. He lasted six or twelve months before he backed off.

She orders another rum and cola, she slowly lights another cigarette. She wants to get this part right.

"Joey," she begins, "takes you all the way down, so far down that the darkness scares you, and there are very few people who can swim on the bottom, and you are down there and it is so cold and dark and you cannot breathe. And he doesn't understand why you can't swim down there. Because that is where he lives."

The band plays,

Take a load off Annie

Bobbie remembers a deal that really pissed her off. A local prosecutor was trying all these drug cases and he'd be in court demanding maximum time for some poor bastard who sold drugs and while he gave his summation he'd be high as a kite on cocaine. She knows all this because they both buy their coke from the same guy, and she has seen the guy's name on the dealer's Rolodex. O'Shay sends her in to memorize the name and number, and every time she goes to make her buy she carefully notes down the name and number, she has a perfect memory, and yet every time she comes back out and O'Shay is waiting in the parking lot, her mind goes blank. This happens time after time. She can't take the pressure, she realizes. But this is where he lives.

The fucking prosecutor, well, he never goes down. Seems the case got rolling and then suddenly this police official weighed in who had a kid in trouble over drugs. He huddled with that prosecutor and fixed things for his kid and then, well, the case on the prosecutor disappeared. He is now a judge.

Bobbie's stories are interrupted by visits to the bathroom. The narcs around her hardly matter. Sometimes, when things get bad, she uses them, gives up a dealer when her habit overwhelms her, gives up the dealer so that her supply will be gone for a spell and she can back off and clean up a bit. She wonders if this is hypocrisy, this periodic biting of the hand that feeds her habit, but no matter, she lives.

The band plays,

Take this gun from my hand

The guitar is screaming now, the bass steady, drums like gunfire. O'Shay dances with yet another woman, he freezes for four beats, then twirls her.

VIKTOR FRANKL, AS HE NEARS his ninetieth year, finally dictates some more recollections. This is hard for him because he thinks ego is forbidden in his case, that those who survive what he has survived have no right to a name but can only report humbly for those who were exterminated.

He speaks in short bursts. In the camp he is leaning over his aged father, who is dying of pulmonary edema and barely able to breathe. He has smuggled in some morphine. He knows the old man, now weak and useless, will be slaughtered before he can die on his own schedule.

So he injects him with a fatal dose.

The son asks, "Do you still have pain?"

"No."

"Do you have any wish?"

"No."

"Do you want to tell me anything?"

"No."

The son leans down and kisses his father. Then he leaves and he knows he will never see his father's face again.

O'SHAY IS ANIMATED NOW as he drives down the old road by the cove. Trees arch over the small lane. The country store, that's where the old men would sit all day and talk. That little pond, "That's where I'd get my crawdads by the bucket load. None of these houses were here then, none."

He hates the changes. The lane leading down to the cove had an old iron bridge "like something out of a novel," and now they've taken it out and put in a simple concrete span. The once empty land is now full of trailers, hovels filled with white trash who keep pit bulls and rusting cars in the front yard, and don't seem to give a damn about the ground that is sacred to Joey O'Shay.

"Human shit," O'Shay mutters.

But the woods are still as thick as when he was a boy.

"I'd walk through these woods," he says, "with a twenty-gauge, hunting rabbits. Or just walking around. The creek goes on and on until it flows into Paradise. And down in the creek bottom possum grapes grow."

The ground is blackjack oaks—too far north for live oaks, which stop at the city an hour or more to the south—and white oaks, cedar, hickory, red oak, native pecan, hognut pecans, scrub elm, blackberries, dewberries, mustang grapes, a tangle of life.

He pulls over. There is nothing but trees and mounds made by fire ants. He gets out and starts walking into the maze. There, yes, there, he has found it. The foundation of that old barn. And over there, yes, right there where the big trees huddle, that's where the man lived all alone with coon dogs. He'd been raised here, spent his childhood hauling rock for his old man, then went off to one of those wars and when he came back, holed up in the woods and just wanted to be alone.

That is when an eleven-year-old boy living alone in the woods got close to him. O'Shay could tell the hermit was strange, but not dangerous.

He had scars all over his body, and the boy never asked about them, but he could see the scars because the hermit wandered the woods almost naked. It would be dark, the boy huddled around his small fire, and suddenly the hermit would come out of the darkness and sit by his fire and not a stitch on him. Sometimes he'd weep.

He taught O'Shay, taught him the history, how the big creek had once flowed and the dam came and the lake rose up and swallowed the creek and that finger came to be called Paradise and that cove became all the world the man needed. Sometimes he'd put on clothes and dig holes or do other paid work, but he hardly needed anything. He ate rabbits, fish, planted a big garden, followed the baying of his hounds as he went for coons in the night. He taught the boy that, too. Where to go for berries, where to cast his line for fish, how to find the rabbits and the coons and anything else in the woods. Hardly a surprise that the boy came to believe that *Huckleberry Finn* was the best of all American books.

Joey O'Shay looks down at the foundation, thorned locust trees growing right out of the concrete. Time passes, but when he raises his eyes all he sees is the forest and then time stands still.

He parks by the lake, gets out, and walks the shore around a point and into the little cove where his creek met the waters. He still has the jon boat he kept on that spit of sand, tied to the big willows. Now the willows are gone, all killed off one year when the lake rose and stayed high for months.

He walks up a small knoll, kicks the dirt with his shoe. As a boy, he found arrowheads here. Behind him the forest almost seethes, a mass of oak stabbed here and there with cedar. O'Shay moves into the trees seeking a giant elm that was the landmark of his childhood. He flounders in the brush and is puzzled at his difficulty in locating it. He finds it smashed on the ground, a three-hundred-year-old tree with a trunk thirty feet in circumference, now shattered by a bolt of lightning. He looks at the water.

"Out there at night, it'd be pitch black, I'd float in my jon boat, stars everywhere overhead, float over there where a spring surges up from the bottom, boils up, and I'd fish. Black as pitch."

He wanders off from the blasted elm and finds pear trees, remnants of a vanished farm. Things change but within the change the thing itself persists. A hermit runs naked at night, coons walk by the shore, a boy hunts the woods, sleeps on the ground, lives alone.

He reaches out, touches the limb of the old pear tree, rubs his fingers on it, eats it with his eyes, sucks it down deep into his lungs. Eagles hunt here, too, and back in the trees, oh, ever so softly, mountain lions pad and seek out the flesh. O'Shay is a boy, going from pond to pond hunting ducks and suddenly he sees a big cat and thinks he is hallucinating. And then two weeks later he sees two lions and the woods grow richer and finer. He finds stumps of small trees, the trunks chiseled off. Beaver.

"In the spring, I'd gig carp. Give 'em to that trash and they'd bury them in their gardens."

He falls silent again. He is back and it has changed and he has changed and he is back and nothing has changed.

He goes down to the marina where he keeps his boat, a twenty-two footer with a 302-cubic-inch V-8 inboard that purrs more than it roars. He casts off, slowly leaves the harbor, and heads into his lake. He goes down the shore a ways to a finer marina, one with fifty-foot yachts moored below hills where mansions stare down. He is here with Gloria, and Gloria beams and the boat leaves the marina and goes out to a strand of islands, humps of sand carpeted with blackjack and cedar and he beaches his boat and they get out and walk the island and the sun is warm and they both love old movies and staying at home and it feels so fine and then that plane comes over and lands at a field out of view, a private field for the rich who huddle above in their mansions and O'Shay cannot stop himself, he is telling her that the plane is his and is bringing in a load and that mansion the plane flew over, well, that is one of his houses and Gloria looks up, eyes fresh and innocent, and beams at her Joey O'Shay. And now she sits in a cage.

Clouds float overhead and the sun hunts through the clouds. Mobs of gulls settle on the dead fish bobbing on the surface and pelicans come in swarms to feed. O'Shay and his boat move across the calm waters and head up the lake, then wheel into yet another cove, a small finger of water with low cliffs rimming the shore. He took Gloria here and in the rock they found claws and teeth left by animals who hunted a vanished world and O'Shay explained to Gloria that this land was once an ocean, once lived deep under the waters, way down where it is dark and cold, but he could tell that she could not understand what he was saying.

Now the cove is silent, absolutely still, with jugs floating where fishermen have left trotlines. The only sound is the thrum of insects and on the shore are little stumps left by the hungers of beaver. A great blue heron slowly flies by overhead.

O'Shay walks the shore.

Night comes, stars return as the clouds move off.

FRANKL IS BEING SHIPPED to another camp along with his wife. The camp is a legend and no one comes back from this camp. So he goes to see his mother, to say goodbye.

He asks, "Please give me your blessing."

His mother cries out, "Yes! Yes! I bless you!"

A week later, she too is sent to the hard camp. And being old and useless, she goes instantly to the gas chamber.

———

O'SHAY CAN TAKE NO MORE LOSSES. The son dying, that was more than he could bear. And now this life he has found on the street is becoming too much. He hits too easily, he hurts too easily. For twenty-five years, he's been out here, alone. And people die around him, his people, and he can't explain to himself what this means or if it means anything. He cannot accept that what he smells and sees and hears and feels means nothing. He cannot accept that he is incapable of believing.

So he flees the city, goes to a trailer in the woods that a friend keeps as a deer camp. And for three days he is alone and he writes out his meaning. He writes a very long prayer.

Father, I come to You late and only when I know I am wounded—severely wounded—as I suppose so many do. I thought I never knew You. This wound I have is of the heart and a killing sort that only the possessor can cure. This possessor is stumbling and not so sure. A street cop I've been for going on the last twenty-five years. The damage was done to the heart from what I have experienced through my eyes and ears.

The cure, I know, I have put off because it is frightening to cover the path where You have been. The years fell away and I am left now a different man that is out and must look deeply back in.

O'Shay scratches out the words about his days in the police academy, his feeling of inferiority because he lacked a college diploma. At graduation, the other cadets gave high-flown reasons for becoming cops. O'Shay was there because he needed a job.

He was there because a kid he'd ran with saw him standing in his mother's yard looking disoriented. The kid was already a cop in a squad car. He'd been hired when he was nineteen and a half years old. They'd talked and laughed and because of their youths, they despised, feared and ran from the cops.

The kid told him it paid pretty good and he thought he'd be a natural. They'd seen the other side, had the jump on most recruits. And the kid was right.

It seems like yesterday, that he gave his half-hearted lie as to why he wanted to be a cop at that first night at the academy with people's parents and loved ones watching.

That hard-eyed, old chief of police, he remembered, seemed to look right through him when it was his turn, and he remembered thinking: he knows I am lying. And the shame he felt for being only one of three in a class of thirty-three with no college degree.

The next time he'd see the hard-eyed chief, when he'd been on about three years, and he was asked to come in to a sub-station by radio on deep nights—FIVE TWENTY TWO, RETURN TO THE SUBSTATION. The desk sergeant told him the chief wanted to talk to him and in the chief's office. He walked into the chief's office, expecting the station chief, a compassionate old cop.

The room of the chief's office lights were out and the chief's huge chair was turned with the chief staring out at a lake and the deep night stars reflecting on it.

The chair spun around and it was the chief, the hard-eyed little man over the whole show and the one who had looked through him that first night before the academy had started.

The chief asked him if he was all right, did he need more days off?

"No."

"Would you do the same thing, if you had to do it again?"

"Yes."

"What are your thoughts, officer?"

"He tried to kill me, chief. I didn't let him."

"Do you need anything?"

"No, sir."

"Officer, most of the recruits from your class have already left us. The first dose of reality blows their idealistic views and their pompous asses. The ones that are left hold life, they are the ones that can take it and survive, that can make a decision in a split second. Those that ponder and analyze, can't take it. It's hard on the soul, son, to kill another human, face to face, so close up and personal. You know, you will never be the same."

Then quietly, the chief said, "I've never been."

Even in the darkness, he saw the hard eyes were now soft. He watched them re-harden as he approached and shook the young cop's hand.

"Son, survive and along the way, you'll learn a lot about people and a lot about yourself. You'll see more in five years than most do in a lifetime. Need anything?"

"No, sir."

"You did good. You survived. Go out there and be careful."

"Thank you, sir."

O'Shay stops writing. He pauses. The meaning must be back there, he is certain it is somewhere near. Outside the scrub forest stretches and the fields, all the elements of a deer camp where men come to kill for fun. He stays in the trailer. He must get this right, sort it out.

The smell, the smell of another person, everything so close, the smell fills the nostrils, that special stamp of another's life, that combination of toil and love and sweat and meals and detergents and soaps, the smell that floods the senses saying this is a life at the exact instant that life is ended, forever.

VIKTOR FRANKL HAD AN AMBITION as a boy when he did not know the camps were being built for him already, being built inside the brains and dreams of other people who lived right around him. He wanted to be a Boy Scout and to own a bicycle. But a war came and these goals became unreachable for him.

He had some triumphs, though, and at ninety, he can still taste their power. Of the hundreds of boys in his district of the city, he alone beat the strongest boy in wrestling.

So many decades later he can feel the sweat of the struggle as he rests on the very lip of death in his old age.

And he had another ambition, this one when he was a young man. He would write a short story. In this story, a person searches frantically for a lost notebook. Finally someone finds it and gives it back to him.

The person who finds it is puzzled by the entries in the notebook, small cryptic entries presented as some kind of code. The writer of the notebook does not explain because the entries are too vital for him.

One is simply a date, July 9th, plus a town and train station. Nothing more.

On that date, a two-year-old slips away from his parents and clambers down onto the tracks. The father finds him and grabs him just before the locomotive rolls into the station.

Viktor Frankl was that boy. The secret entry is his history. In the story, the reason for the code, for the secret history, is never given.

O'SHAY HAS SHAKEN OFF the party, dismissed six hours of dance and alcohol and fellow cops.

Night has fallen, the moon is down and the stars roar overhead. He is summoning up his feelings from some well he sealed off long ago.

That smirk, I hate that smirk, I've seen it before. You come back to the office after just having hurt someone or put them away forever and the supervisor is there and he's makin' $40,000 a year more than you do and he's got the same damn smirk on his face. And he doesn't know, he's never been there, hell, the guy I just took out was a better man than he is.

And then when he thinks of the pain he has caused he knows even such thoughts are a sin.

THEY'RE KILLING HIS PEOPLE, and this man knows it. He is Rabbi Yehuda Low of Prague and he is an ancestor of Viktor Frankl. Rabbi Low finds the answer in a dream: create a being to frighten away the killers.

He calls for his wife and a student to help fashion the being from earth, water, fire, and air. For seven days, the rabbi and his wife and the student ready their souls for the work and then they go to the river and get clay. The creature they make is ten feet tall with features somewhat like those of people. The rabbi's wife circles the lifeless beast seven times and says a certain phrase. The student circles seven times and says another phrase the rabbi teaches him. Fingernails start growing, hair sprouts and the body is suddenly a mass of fur.

Now the three of them recite Genesis 2:7: "And God breathed into his nostrils the breath of life and man became a living soul."

The beast opens his eyes. He is called a Golem.

The rabbi names him Joseph, tells him his duty is to protect people from evil. The Golem nods.

He lives in the rabbi's house, sits right there in a corner, wears clothes around the house. He does nothing.

His work comes. The killers return to the ghetto for their

prey. A child has been murdered and the crowd blames the rabbi's people. The Golem pushes his huge body into the throng, plucks out a woman, and she soon confesses to having killed the child.

The danger passes, the killers cease coming, the town council even passes a decree outlawing such actions. The Golem's work is done. The rabbi, his wife, and the student return him to the quiet he came from by circling him and uttering that certain phrase. His still body is wrapped in a prayer shawl and rests in the temple.

He becomes folklore.

And he seems a lot like the Evil Creature that watches over Joey O'Shay, the thing that hovers over him as he lies in the back room dying at age four.

VIKTOR FRANKL AT AGE NINETY remembers when he first came to grips with death. He was four years old.

It was evening, the boy had almost fallen asleep when the thought came to him: one day he too would die. He did not fear dying. Not then, not later. He feared something he could not say properly for years: "whether the transitory nature of life might destroy its meaning."

That life ends and its meaning vanishes, he struggles with this possibility for years and years. Then he decides that death is what gives life meaning, that the fleeting moments humans call life are not emptied of meaning by death because "nothing from the past is irretrievably lost. Everything is irretrievably stored."

THE PAUSE IS OVER. He picks up the pen.

He was there because of a disturbance call.

Ah, he pauses again, that third person, it feels right, gives

him some distance from where he is heading. He was there because a man with a gun had threatened to kill his wife.

*All had been quiet in that early morning hour. He still
remembered the way the night smelled—dank and still.*

Ah, details hardly matter, what happens matters.

*He felt in his guts that evil lurked in that dark old house
and he could smell it. He thought of his little boy, pulled out
his photo with the four-leaf clover he had found while in the
Army and put it in the picture beside the little boy. And he
remembered to be determined not to lose, knowing it was
coming.*

*The man in the hat, pitch black bedroom, he came toward
the lighted hall.*

The man came with a .357 in hand.

He was not afraid. Timed it out slowly. He felt strong.

The man was wild-eyed and large.

He grabbed the man's hand as soon as he whirled around.

The man tried to shoot him in the head.

The man's gun went off.

*The officer put two in the man's torso at point blank
range. The impact sent the man hard backward and that
death grip took him on top of the man.*

Their eyes met, their time slowed down.

*He saw the man's eyes fade from rage to surprise, blood
came in a gusher towards him...from his mouth.*

Upon him who had put him there.

*The man's eyes faded and were lost in emptiness. The
man's limp hand still held the gun.*

*The gun was pushed away. And he stood up in that dim
yellow light of the hall and thought...*

"Hell," he said, "I had to kill the man."

He turned and looked at the human shell he had left in the hallway.

He walked alone and the man's wife stood in the yard. He talked to her alone and told her he was sorry, her husband had tried to shoot him and he had killed him.

Her blank look.

O'Shay pauses again, the pen stops moving.
He resumes.

And I wonder where You were, Father. Why do things happen? Like this? What ridiculous set of circumstances made me and this man—who I had never seen or spoke to— engage in an act so grotesque? And final.

He has to stop again. That dank air, the dim light in the hallway, the woman in the yard, him walking past, that slight pause as he says, "I'm sorry."

He sat alone in a cold office and explained his actions. He took out his son's picture and looked at it. He would engage in a sad...

He has to stop again. The woods are outside, the woods are all right, the woods are. He is in the trailer, alone. He must get this down and get it right.

Were You there, Father? In that dark hallway, Father?

SHE IS HIS BRIDE. And she is beautiful. But Frankl marries her because she tends to him and he needs the comfort she brings to his life. Once when he had to rush off to deal with an emergency at work, she held his lunch ready until his return. This act touched him deeply and made him want her.

Nine months after the wedding, he and his wife are in a concentration camp. He knows he will be sent east to a harsh camp and that if she tries to go with him, it will likely prove fatal. So he realizes he must stop her from this act of love. He fails.

During the few minutes they have together in the new camp, minutes he senses will be their last together, he tries to make some gesture of his love. She has already smashed an alarm clock she has brought lest the camp guards seize and enjoy it.

The moment comes when the men and women are sorted out.

He says firmly to his bride, "Tilly, stay alive at any price. Do you hear? At any price!"

His meaning is clear: sell yourself sexually, if necessary, in order to stay alive. He realizes he is giving her license to survive so that he might lessen his guilt at being incapable of saving her.

Like all wars, this one finally ends and Frankl is walking across a field when he runs into a Dutch laborer who has also just been liberated. As they talk, he notices the laborer keeps playing with something in his hand.

"What do you have there?"

The man shows him a small golden globe with the oceans a blue enamel. The equator is indicated by a gold band and the inscription reads: "The whole world turns on love."

Frankl had given such a piece to his wife and at the time he had been told only two such items existed in his city. He buys the small pendant from the Dutch laborer, though it is slightly dented.

Later, on the first morning he spends in his city after liberation, he learns Tilly survived the war but died in the weeks following from disease and starvation. This is in August when the air is fair and the sun warm and golden.

THE PEN. THERE. PICK IT UP. Become he, the third person, that cold object safe and sound and out there, the person these things happen to, the person kept distant from yourself and your pain. But the distance is vanishing.

Forgive me. It is said that in nearly all cultures that as one passes from life to death the kaleidoscope of one's life is seen in an unexplainable panorama. The good bad happy sad, all that is significant, those moments that carved you from the formable clay to the finished product. The clay can be new and innocent as angels when taken. It can be hard and strong when taken. It can be old and deteriorated when taken.

The clay is formed by chance and free will by the artist.

The clay is the artist, sculptor and his medium is himself and where chance has thrown him or her.

The clay's original form has You fully in it, Father, for we are all born innocent babes in this harsh free-willed world. The clay forms gradually and in the end when the clay image has no life what is left can be sad and selfish, knowable and giving, according to the sculptor's choices. Damages in the finished product in the end are most critically judged by the sculptor himself.

Can they look upon their work as ugly and harsh upon the soul or can they see meaning and beauty in their final work?

Do You guide us silently, Father, in Your artistic endeavor?

Forgive me, Father.

You've seen into the kaleidoscope and the images have not blended yet, for it is not the end. The images are there though, images made by choice and by free will and chance. The images that break the heart and haunt the soul.

How an old patrol sergeant told him as a rookie: they will come and sit on the edge of your bed at night, the ghosts. They exist in no timetable and follow no chronology.

They're part of him and though neurologist and scientist may explain away what memory is, in his damaged heart he knows that what life has absorbed is life's significant offerings—to be ignored or used for good or used for bad. He knows this but they are no less frightening. Because he also knows that within all men still lies at least a piece of the original clay that misses where it came from.

He stops again. He knows what comes next.

Though we have that and miss that, why do we find ways to avoid that?

Forgive me, Father.

Why is that, Father?

Forgive me, Father.

The old woman is horribly burned. She has somehow and for some reason dragged an old mattress out of the burning apartment in the projects. The young patrolman hears her moaning out by a vacant lot. Sadly alone, she will die this cruel death in a junk-infested lot dragging a twin-sized bed mattress to lay on.

The young man is wild eyed and screaming. He has completely covered himself in gasoline and holds a cigarette lighter.

His mother begs him in Spanish to stop. She begs the young patrolman to grab him. He's mentally ill, she yells.

The summer night itself seems hot enough to ignite him.

The old veteran shows up.

The teenager is more disturbed, growing threatening.

The young patrolman is terrified. Dying in fire is horrible.

The old vet throws his flashlight, hitting the boy in the chest, knocking him backwards.

They grab him. In slow motion, the young patrolman sees the kid trying to strike the lighter. It doesn't light. They get it away. All are now covered with gasoline.

The vet and the young patrolman talk later. The vet tried the lighter again. It worked.

The boy shoots himself in the head two months later.

It's cold and rainy. He cruises the projects in his patrol car and notices movement in one of the cement overhangs over the project doorway.

It's a ten-year-old kid.

He shines his light. The kid comes down, saying he slipped out of his bedroom, which is on the second floor.

He asks the kid, why?

The kid instantly says, he doesn't like it when he can hear his mom and her customers in the next room.

The young patrolman drives him to a 7-Eleven and the kid gets a chili dog.

He returns with the kid to the apartment, beats on the door, the trick and an alcoholic mom come to the door pissed off.

The trick talks ugly and is kicked in the balls by the young patrolman and sent puking into the parking lot. Mom is threatened by the young patrolman, the woman admits the kid was some white trick's kid, says only God knows which.

The kid gets out of patrol car, tells Mom it will be okay. They disappear into the three a.m. darkness of their apartment.

O'SHAY IS FASTIDIOUS. He keeps his boat out of the water. A hydraulic system lifts it up from the corruption of the lake. He buttons down the canopy and this seals off the cockpit, even though the slip itself is roofed and secure from weather. The boat is immaculate, brushed, washed, scrubbed, every surface shining and clean.

He must have order.

In the living room of his house, a stained-glass window faces the street. A caravel is under full sail in the glass with orange sails riding above a green sea. The ship is canted from the force of the wind and storming across big waters.

None of these tactics makes O'Shay safe.

As a boy, he had this metal barrel and he would get in the barrel and roll down the hill behind the house, down the hill to the bottom where a magical mulberry tree grew, a tree festooned with butterflies and wonder. When he lay dying as a boy in the back room, the Evil Creature came loping up from under this tree.

One day he is in his barrel and rolling down the hill. He comes to rest at the bottom in the sun and his legs are jammed and he cannot get out. He is stuck for hours, the metal gets hot, his body burns where he touches the barrel. His mind seems to go.

He sees two small green creatures, like elves, and they have cloven hooves. They are named Jack and Jay and as the heat builds and his flesh burns, they seem to dance just outside the opening of the barrel.

Eventually, after many hours, his father finds him and wrenches him free. He is laid up for a spell with the burns.

He never doubts the elves.

HE WAVERS AS HE SITS in the trailer alone trying to write. He hesitates, wonders what to put down on paper. The composition seems to have its own structure and he is simply learning this structure by writing it down.

He goes to the hospital. The Irish nurse, the one with the accent, has gotten to know him. On the deep night shifts, so many calls he answers there because his beat is adjacent to the hospital where the poor, savage, and ravaged must go.

She tells him it is a little boy, horribly abused and neglected. So prepare.

There is no preparation for what he sees. The child is obviously severely retarded. Its lips are gone, its fingers are all bloody nubs with bones exposed. It is covered by cuts and sores. It appears emaciated and gaunt, like an old man.

It's four.

He sees movement for the first time and thinks he is hallucinating. He realizes it is maggots all over the child. He has to hang on to the child's gurney not to fall over in a faint.

The nurse explains the kid was tied up and left and that the maggots ate the decayed flesh and kept him alive.

He feels like ice is in his blood. The stupid, loudmouthed mother is raising hell in the hall. He wants to take his baton and beat her till the brains run out.

He handcuffs her, arrests her for her child neglect and takes her to jail.

He leaves the jail, pulls the squad car to the river levee and tries to fill his lungs with clean air. Only the deep smell of befouled river meets his nose. An old patrolman once told him to go home and let the hot shower take it all down the drain. A long, hot shower.

But the picture of the kid never gets washed away. Thirty years later the kid is still there.

He and his partner answer the call at three a.m. to meet the teenager. She tells them she is tired of her father raping her. She is thirteen-years-old. She has been raped since she is six years old. Her mother knows it. She says, he just raped her.

They go back to her home. There are crucifixes and photos of Jesus in the living room. The father is dirty and pathetic. His breath is rotted. He doesn't deny it. Says he was drunk.

Can a person get that drunk?

They take him to the squad car and the veteran cop punches the guy twice in the gut.

"That's for the little kid, you cocksucker."

The man cries.

Six months later, they see him buying beer. They jack him up. The girl lives with her aunt. He gets probation.

———

A TALL WALL OF WOOD surrounds the backyard of Joey O'Shay and yellow honeysuckle flourishes on the surface of the weathered wood. Trees bud out, spring is near and moves silently through the air like a drug. Grackles hunt the lawn, filthy fucking grackles that remind him of a society gone mad, but they come and that is part of life, too.

The sky is slate gray and yet light seems to walk across the yard and grass has fire burning within it from this filtered light. Toward the back, the winter garden creaks on, lettuce and spinach having the final days before they bolt to flower and go bitter to the tongue. A small metal sign says, DAD'S GARDEN.

The yard is normal, as is the house, one of a string of bungalows that huddle in the old neighborhood of the city. They are stucco and brick and all the lawns are neat, the trees mature, the driveways swept clean.

Inside the French doors, where the black sofa sits in the back room, there is a large sack of peanuts in the shell. Squirrels leap from limb to limb in the tree. Finally one descends, crossing the lawn, creeps under the ramada by the doors, and scratches ever so softly on the glass.

O'Shay rises, gets a fistful of peanuts, opens the door, and spreads them out. Geezo comes near, looks up briefly, and begins to feed.

He is working murders, ones done by a syndicate. He feels good about this fact. Nothing justifies murder, he thinks. He is meeting with an agent, a big guy who has worked the border and other places. The guy comes to his office and O'Shay notices him looking at his voodoo altar. He is struck by the decorum: the guy knows enough not to pick up the various skulls and items as if it were some curio shop.

He simply stands there and smiles at it for a good long while.

Then he turns to O'Shay and says, "You're a conjure man. Louisiana?"

O'Shay says, "No. Paradise Cove."

The big agent smiles again and says almost to himself, "The conjure man of paradise."

O'Shay replies, softly, "I fucking wish, kid."

HE PAUSES, GETS HIS BEARING back, and determines to keep writing.

He must tick off his dead.

Why the young ones, Father?
Are they not the innocent?
Forgive me for asking, Father.

A Native American on patrol, a friend who lived a blind and beautiful faith. That January night he pulls a car over, the shots ring out, he dies.

A deep, vindictive hate and need for revenge was born in him that night. His killers were eventually caught. An unplanned, needless murder. He'd stopped them to tell them to turn their lights on. He died trying to be kind.

The huge black cop, kind and religious. He answers a disturbance call, a kid shoots him with a .22.

He told his partner he was dying, to tell his family that he loved them and that he loved his partner, too. He bled to death.

The next one was white, pulled over a stolen car.

And he believed in You, Father, with a deep devotion.

The fatal bullet goes through his eye.
The roll call continues.

He knows the real victory, on the whole, is lost, truly lost.

In thirty years, forty-three of his brothers in arms have died on the streets in this city of the United States of America.

Why is it, Father, that our best are truly taken away or unjustly treated?

Is it dumb to really believe or to really care?

Forgive me for that question, Father.

There is so much that has taken its toll on a man's heart.

IN THE CAMP, VIKTOR FRANKL tries to distance himself from what he is experiencing. A trick of the mind. He is marching one morning, his feet swollen and festering, he is hungry, he is cold. He feels his life is hopeless.

So he imagines making something of it. Yes. He sees himself at a lectern, the hall warm, lights bright. He looks down at the paper he is going to deliver: "Psychotherapeutic Experiences in a Concentration Camp." He reports in this imaginary lecture the very things he is living through as he marches in the cold to work that morning.

He calls this self-distancing and he is convinced this fantasy work keeps him alive.

This and the other task, a book he will write when he survives, the book whose manuscript was taken from him and destroyed. He must reconstruct this book because unless he does, unless he makes sense of what he has experienced, life will have no meaning.

And then, he will die. Or surely should die.

HE IS LOSING IT, that nice strand of narrative. Things are shattering within him as he pens the words. Bitterness rises in him as he thinks of his dead and of those who make careers off the bones of his dead.

He remembers:

The five kids splattered and decapitated on the road from an explosive wreck.

The shootout. Making entry, two officers shot approaching the house. One, his head burst in back and bleeding profusely, draining blood, shot in the lower stomach.

The other shot.

Going back into the darkness of the hole, finding the suspect lying on his face.

The officers whispering how to tell the kids and wives without terrifying them.

He heard his partner once remark, when they were trying to drink away the fact that they dealt with what they dealt with, was really broken dreams and broken lives and lacked the correct tools to repair them or themselves or the people, that there are people that know, and those that never will. We know each other when we see each other—these people that know—usually without saying a word, we know each other mostly in the eyes.

They do know each other when they see each other.

They feel each other's atmosphere and eyes and it gives them away.

He hesitates again. He feels awkward. The trailer suddenly seems to be closing in on him. But he cannot stop himself from this insistence that he is not alone, that there are others, that he is not the first nor the last.

He knows the others can know but often don't by free will. He has noticed in his aging that greedy, selfish people seem to get what they want on this earth. And he has also observed that it is never enough. They seek more and more. They are hollow, sad, and empty. Even they know things that they seek cannot give them a good solid conversation from the heart.

He knows that it is easier to turn away from what is meaningful and take the road of the prurient and self interest. It is easier to avoid the tragedies of life—isolation, hate, cruelty, selfishness—than to care and help. He knows that to create, cure, guide, tell the truth, and be compassionate are not accepted as important in a large portion of humanity.

To give of the heart can take a tragic toll on the individual shell but it strengthens the soul.

Humanity killed your only begotten son, Father, did it not, Father?

Forgive me, Father.

O'SHAY TAKES OUT HIS powerboat and glides across the lake.

He has his two boys with him, now grown and strong and sound. The lost son is a whisper. Just as the long manuscript he wrote those three days in the trailer at the deer camp is present and yet not mentioned. Nor even known to his sons.

But there are things they have sensed wash across their father's life and resented, and endured. And now they are on the lake where the sky is blue, the sun is warm, and islands and coves beckon. This is the geography of their childhood, too.

They put out trotlines in the deep cut near the islands and get five catfish. The blue cats come out of the water a baby blue. A big channel cat has a spotted golden hue and O'Shay wonders if the rich color is a by-product of the alga bloom that is murdering so many other fish.

The boat takes them out of time. The city ceases to exist. The sun caresses the skin.

IN THE TRAILER, as the days crawl past, O'Shay finds he cannot stop what he has begun. Something is boiling out of him and he must endure until this need is finished and on the page.

The night air has the first hint of morning. Unlike the summer air, the slightest touch of cooling breeze washes past him like a giant sigh. He has faced the abyss and come to terms. He knows he has been very damaged but not broken.

He comes to terms with things he has never really felt comfortable with. He has come to terms with himself. From the abyss, the tragedies have not gone unheeded, for they are part of life. They make the wonderful things that are all around us seem so much more wonderful. To forget that from time to time is simply human. To live in isolation, ignore the tragedies, and not appreciate the wonderful, beautiful things and people is to not seek the meaning of life.

It is for mankind to suffer in varying degrees but it is his free will to take the hand that is dealt to him and find significance and beauty. We must remember people who have survived tragedies so immense as to seem inconceivable. And they not only survive but reach out and touch thousands with kindness and humanitarian efforts.

So, there are angels amongst us, Father, though they have no wings.

Why not a few more, Father?

Forgive me for asking, Father.

The dawn is breaking slowly on this earth. The abyss is gone.

What broke his heart and made it bleed will not heal his soul.

He knows he will never be the average person, no one that survives wars of any kind is.

They carry the ugliness and lies of what the war was fought for.

They know that there is not but a handful of bad guys on either side.

Most of them are just some mother's son trying to make some sense of the ugliness for survival. They also know that

very few of the really bad guys fight the war. But the cruel
and selfish little men with cruel and selfish little reasons and
the mothers' sons fight for all the wrong reasons. The ghetto
kid sells dope to get out of the ghetto and never does. The
detective stalks them to end the flow of the poison to the
community and he can't.

He knows that he is lucky, too, lucky not to be killed or
maimed worse than just a leaking heart.

Did You ever look out for me, Father?

Forgive me for asking, Father.

Lucky because of his three sons. The oldest seeks himself
and went to Bosnia and saw the cruelty and ridiculousness
of war. He flies high in the sky each day now above the chaos
below. He writes beautiful sad songs and beautiful poetic
songs, he cares about life and nature. He who left the most
wonderful memories of the extra point pass to win the big
game, the center field homerun...jars of cicadas...the fact
that this young man's existence never let the man give up in
a gunfight or fight.

He knows his son knows and his pride is unparalleled. The
youngest who has always been so kind and gentle, though
he was so big and strong, always compassionate to the
small, slow, and less fortunate, playing like a gladiator on the
football field, sweating, pumping iron in the heat. Never a B,
always the top of the class. Memories of a chubby little guy
running on the beach with me, nearly falling off the top of
a houseboat. Now a powerful young man. The fact of his
existence would never let him give up in a fight. He knows
his son knows that his pride in him is unparalleled.

His little son whose baby's last breath as he lay dying
made him hurt so that he wanted to die. The little boy died—
and Rocky, you are in a better place and you made him seek
the meaning of life. Without your sacrifice, it would not have

occurred to him to look. He knows you are with him and you are an angel.

O'Shay cannot keep going. He stops, he has shortness of breath. He has ticked off his assets, he has tried his best to keep an honest accounting of his debts. But this last part, this costs too much.

Did You hold him for me, Father?
Did You rock him gently?
Did You rock my little boy? Gently?
Forgive me.

HE IS BACK FROM THE CAMPS but Frankl is not yet really back. He is practicing medicine but he is not well. He has things inside him and he does not know what to do about these things. One day, a year after being freed from the wire, he scratches out a poem on a prescription pad. He is not a poet, but still the lines come to him with a fury.

You weigh on me, you whom I lost in death.
You've given me the silent charge to live for you;
So it is for me now to erase the debt of your extermination.
Until I know with each ray of sun
You wish to warm me and meet me;
Until I see that in each blossoming tree
There's someone dead who wants to greet me;
Until I hear that every bird's song is your voice
Sounding out to bless me and perhaps to say
That you forgive me that I live.

HE IS WRITING faster and faster.

To truly look at the stars and sunset and sunrise, the clear ocean and sound of the surf, the rustling of leaves in the

wind, the smell of rain coming and the softness of snow,
these are but a part of the immense gratitude we feel to have
had this particular life.

You were there, weren't You, Father?

Forgive me, Father, I know.

They say the secret of life was hidden from man at the
first in the ocean. He would have trouble finding it there in its
vastness. Don't hide it in space, very little chance he could
find it there. So he hid it within man's heart which is the
hardest place for man to find. He must look deeply within
himself.

Father, I came to You late and only when I knew I wanted
it so badly as I do. I thought I never knew You.

The sun is risen. Last night's frightening abyss is gone. The
scene of the past tragedies is cleansed by your dawn. It seems
that I've been trying to find life's meaning as my real goal.
Making it all the more ironic that my wound is of the heart
and of my soul. Some people, they find it so hard to give.

Father, I gave but oh, how I have lived.

Some say I sought tragedy and its haunting sadness is
how I now pay.

I know You now, Father.

Forgive me. You shadowed me every step of the way.

VIKTOR FRANKL GOES INTO THE ROOM, a drab room with no heat.
The war has erased comfort from his world. Cardboard covers
the broken windows. He dictates to three stenographers who
work shifts. He paces. He sits in a chair. Sometimes he sobs.

When he finishes, he has re-created the book he lost in the
camps. But he is not yet finished, he discovers.

So he starts another book and for nine days he dictates to three
shifts of stenographers. This book is accepted by a publisher and
Frankl has only one demand: his name cannot be on the book.

He wants anonymity because the book is about the camps and he feels no one has a right to speak for the dead and claim some kind of authorship. While the book is on the press, friends and the publisher finally wear him down. The first edition has no author on the cover but does disclose his name on the title page.

The thin little book, *Man's Search for Meaning,* is finally published in the United States. And years later, it comes into the hands of Joey O'Shay. And Joey O'Shay eventually goes to a trailer in the woods and tries to do what a man from the death camps found possible.

HE WAS HUNTING ONCE in the forest by the lake. He'd gotten his draft notice, his wife was carrying his child. And so he went into the trees. He didn't want to kill anything, he wanted to be with things and the hunt took him into this other world. Then the snow came down, soft, white, pure snow. The forest filled with silence, tons of silence, everywhere.

Snow hanging on the cedars.

He stops, the flakes brush his cheeks. And in that moment, sky gray, world turning white and pure before his eyes, he suddenly knows everything will be all right.

He knows he will have a child. He knows the child will be a boy. Snow falling, the draft breathing down his neck. But it will be all right.

He walks the silence of the woods, white everywhere.

And he knows.

There are these moments, the atmosphere can change, ghosts can show up.

And it will be all right.

HE IS IN THE KITCHEN AT NIGHT. It is Christmas and one of his sons is home and his wife is home. He glances down the hall and suddenly sees an old woman in a gray dress move past the doorway to the dining room. What the hell? he thinks.

He goes cautiously into the room and then the adjacent living room. He sees nothing but feels someone. He calls for his wife and son and they see nothing but feel something. The room feels oddly cool even though the fireplace is going. O'Shay feels something like electricity in the air, something very happy. He feels as if he is going to weep.

After a few minutes, the feeling passes and they are alone. And O'Shay feels very sad as the presence leaves his home and he is there alone, once again.

Joey O'Shay is certain the woman he saw is his dead grandmother and that she came back to feel the holiday and enjoy the tree.

His mother sees things other people cannot see. She dreamed her grandson was going to die and how he was going to die. And then the boy died. When O'Shay was a boy, his mother took him to the store and she saw a neighbor she did not like. So they left. And when they got home, they learned the neighbor had been dead two days.

Ghosts, spirits, he does not know what to call these things. Nor does he feel the need to explain these things. All through the days and the nights a dead one-year-old boy has been with him, a son staring him in the face from the grave.

He feels in his heart that those who go early, who are not born right, who come into the world doomed like his lost son, that they go back instantly into the ferment from which they came.

Because of his death, I kind of took a journey and disgusting as it was, I tried to end up learning...I was seeking meaning in his death. It could not be meaningless. Yet a part of me thought it might be meaningless. Gradually,

I evolved to a point that the things I was doing…led me to
magical people, good people…there is no way to compare
my pussy-assed life to what others have suffered and
suffering has dignity and in that dignity…I don't know if
Christ was a man or the son of God but I do know he never
wrote anything about himself and yet he had such an
amazing aura and energy…he was saying this has meaning,
help people, be good, listen to your heart…and we have
created drugs to wall off our hearts from the magnificence of
the day, the colors, the sounds.

HE FRIES LEAN SAUSAGE, grills potatoes, cooks eggs, grates cheese, chops some onion. He makes twenty-five breakfast tacos. Packs tortillas, jalapeños, and hot sauce. Brews good Guatemalan coffee and fills two thermoses. Slices a honeydew, a cantaloupe, some oranges and limes.

Joey O'Shay meets the lawyer at the jail and both enter to have a sit-down with Gordo who is going away forever and knows it, and knows that nothing he says can help him.

But good coffee, some real food, a meal on his way to nowhere, that matters. He gives up everything to O'Shay, the string of murders by his nephew Johnny Boy, every motherfucking thing.

O'Shay thinks, Hell, he isn't such a bad dude.

Then he drops by a party Bobbie is tossing and she looks scared. And she is not well and he can see this fact in her skin and in her eyes.

He gets home and there is a message that a young prosecutor, a good one, slit his wrists. He'd been demoted, took it hard. O'Shay thinks,

…I hope he is in heaven. The street gives and takes…
supple then savage…beautiful…then grotesque. The street

*is free will and what we run to and run from. It is lessons lost
and lessons learned. It is for always and forever...*

He thinks he is a little drunk.

HIS FACE IS BLANK, the eyes deny any entry. There is a silent speech
running inside his skull.

*...I am essentially going to turn around and say that
everything that I did was wrong. I am not doing it to be
some kind of a goddamn martyr. I am doing it because
I believe it is true.*

I'm laying it down because it is wrong.

*It has become a goddamn business, a corporation. The
government is spending billions.*

*You can't just turn your head away from it. You've got to
make it legal. And control it. You can never keep people from
synthetically trying to reach out to God. All people do with
dope is a veiled search for some kind of higher being.*

*I know Gloria was essentially a good-hearted person who
chose an easy out. Had I left her alone she would probably
still be dealing dope for money. But that is between her and
God. She ain't no fucking saint. But she was dirt poor when
she was a kid. She smelled good, nothing about her was
fucked up, she was polite.*

*She didn't have the stink of death about her. I feel guilty
because I think the mutual attraction and respect was the
cause of her ultimate annihilation. I was like a vampire.
I came to her in the night, was her friend, and then sucked
the life out of her...*

COME, SMELL, LOOK, LISTEN. There, the sea spreads out and the
sky is so blue, that intense blue that aches with promise and the
man named Joey O'Shay is vanishing, the last name is already

gone, erased and cast aside, a name never to be used again on the streets, and now this other Joey tastes that blue, smells the salt air, listens to gulls' cries, and thinks he cannot face this blue in the sky, this slap of water on the beach that is called an ocean, cannot face these things without a drink to deaden himself and to fashion a new Paradise Cove between his mind and heart and this unbearable world.

He watches a woman walk down the beach with her granddaughter trailing by her side and suddenly he sees Gloria, but at this moment, Gloria is sitting in a cell far from this Central American country. And just a week or so ago she was asked once again to explain this one entry in her phone records, this man named Joey O'Shay, and she would not do it, she would not give him up, not say a word about this person who now has ceased to exist. She asked the agent, "Did you ever get Joey O'Shay?" and the agent said, "No, he got away," and Gloria gave off this soft smile, a flicker across her face.

Her children still call that old number, the one that belonged to a man who said he was Joey O'Shay and the phone rings in that cold federal office, the strange office where a voodoo altar rears up with candles and skulls, the office where a coffin waits for those who have been bad, and these calls are never answered, the instruction has been given by O'Shay never to pick up that phone.

He has fled to this beach in order to heal or rest or do something, to shake the sense of impending death from his body, to flush his cells of toxins, to erase scents from his mind, to join a greater world, one where boats sail and porpoises skip along on the wave lashing off the prow, yes, he does this, sails into the sea and feels a rush of wonder wash over his life and this feels good and fine and he thinks he is coming around, getting past all these things he has dragged rattling behind him for decades, he thinks as the weeks roll past that he will become a new person, one safe from the case he has won, safe from the lives he has ruined, safe from the smile on Gloria's face as she goes to that final meeting, her body sculpted, her hair and nails done, her perfume a soft mist that would snare any man's heart, yes, he must erase Gloria also, and he must forget that during the last leg of the case one of Irma's couriers died when a condom full of heroin burst inside her, and how Irma's people sliced her open and then cut her intestines out and slowly squeezed them, yard after yard of them, squeezed them until the very last condom of heroin fell out on the table. And then burned the body and cleaned up the mess.

All this must go, everything has to be erased and he is doing well, the blue of the sky can be handled with a drink and then the woman with the granddaughter walks down the beach and he sees Gloria's face on the woman, can suddenly feel his nostrils fill with her scent and he thinks this may take more time, take more days sailing on the waters. And then the night comes and the stars are out, blazing stars that can never be seen in cities and he sits on a balcony and knows these stars cannot be tolerated or endured without more alcohol.

He wants out of his life because he is certain his life is killing him.

But he knows that if he leaves his life he will die, he has never found the path, the way, the sentence that can massage this truth and save him. If he stays, he will die. If he leaves, he must die because there will no longer be a reason for him to live.

There are rules about stories, rules that say stories have a beginning, a middle, and an ending, that stories begin with questions, with mysteries and then answer these questions and solve these mysteries and these rules guarantee that stories satisfy and calm and ease the burden of being alive. The cop faces the dark alley and fills it with light, the agent meets the dazzling woman, succumbs ever so briefly to her charms but then rises above such temptations and destroys the woman as a testament to the eternal truths and values of this world. The bad guys lose in the end. The good guys win. The world returns to being a safe place. The page turns, the case is closed, the threat ends.

The stories that follow the rules always lie, each and every one of them. Still the rules must be observed. He is on the beach, the sun warms during the day, the sky is so blue, and at night the stars dance and a low hum comes off these stars and sings to the world below. All the women smell of coconut from lotions that coat their soft skin and food wafts through the air from the small stalls studding the beach and the drinks are always cold and at dawn the gulls cry and out to sea the porpoises wait in schools. Such are the rules. But he must get another drink as night falls down, must have one immediately in order to protect himself from the invasion of the stars.

God knows he has tried to protect himself, he has fought the danger of the stars and of beauty and of love. And he has fought to risk that danger. He stands on the balcony looking out to sea, drink in his hand, memories of Gloria floating in and out of his mind, and behind him in the room, she is sleeping, her dark skin rich with the scent of coconut lotion, and she did come back and he did take his chance and face the danger, that woman in the white sundress, the one who said they could not love, that the world would not permit their love, the woman who spent that perfect night with him in the tower, her skin then a cloud of mint and vanilla caressing O'Shay's senses, the one who drove away from him because he was white, she came back and that rare chance at life that almost never comes, he took that chance and she has kept him alive when he thought surely he would die and rot inside. Now they are here, on the beach, so he can heal, so he can forget what he has done, erase the case, leave Gloria to her cell and long years.

And he knows the thing called the case, the destruction of Gloria, the whole matter is very simple and very hard.

He did not betray the woman in the room who once wore that white sundress on a lazy fall afternoon while red and yellow leaves fell like dreams onto the ground.

He did not betray Gloria who played out her hand in the only hard world she ever knew.

He betrayed himself.

He enforced a law he no longer believes in.

Life has become what he always desired and now the desire eats him cell by cell. And memories come up and hammer with hurricane winds. The phone rings back in that cold office and it is Q, the man who had the tape of Popsicle's last night, the man who suffocates her, Q calling from prison and saying, *"You gotta do something for me, the Mexicans are killin' niggers in here."* O'Shay says, *"You killed that little girl and you should suffer,"* and slams down the phone and Jesus the stars are threatening, staring down at him as the waves softly lap the beach of endless night and he sips his drink and listens to his mind, *I always enjoyed my job, it was a place I could learn but the whole picture made me feel pain in my heart and I thought I was doing the greater good and now I feel like a whore working for a pimp.* He sips that drink, takes in the stars, the same stars that claimed him as a boy at a cove called Paradise, cold and distant and clean and honest and blank and angry as the face of God and *this has to end but I can't function anyplace but where I am at and I can't even work homicide and I threatened a sergeant the other day* and he is leaving, he insists he is leaving but all that seems left of him is what he does.

Cosima snaps at him, "You in love with Gloria?"

"Yeah, I'm going to marry her when she gets out of the pen."

"Slut."

This rumbles in his head as he stares at the blackness of the night sea and he wants to laugh and make light of it and what the hell, business is business, and still the winds come out of his past and when Cosima gets her payment for doing Gloria, suddenly she blurts out, "I'm a bad person. I sold my friend," and Joey nods because he must agree, he sells people himself.

The drink, he must keep refilling his drink, leaving his balcony, going into his room, filling that drink and then returning and watching the darkness of the ocean and *must have a drink to face the beauty, to keep it at bay, because when you join the beauty they take you away* and he can't talk to anyone about these things, ah, the rustle of palms in the night, that gentle lapping of the waves, salt in the air and somewhere out there porpoises moving with perfect grace and *you learn, you find the meaning, might take a thousand years and a hundred lives and I feel like I've been chasing things since I realized I didn't belong here and the old man is standing in that house in boxer shorts yelling at the old lady and down near the creek in back the Evil Creature and I must have slipped through a crack and wound up in that city* and the case is closed, the case of some kind of lifetime, tally the numbers and Gloria was moving at least 200 kilos of pure heroin a year, an amount worth $100 million on the street and Irma was moving at least 200 kilos of pure heroin a year and so the numbers say this was a $200 million bust. Or the deal was worth $5 million a week street value and that means $260 million a year and there were two separate groups vying for the deal and so that means more than half a billion a year. Or it means nothing because Joey O'Shay was not really a dealer and the hundreds of millions of dollars never really changed hands. Depends on how you want the story to go. But one thing is certain: dozens of people are ruined and sit in cells and now O'Shay is free and watching the stars at night while he clutches that drink to protect him from the beauty *because if you go to the beauty, you are at peace and then you no longer can do the deals, do the work and then you no longer have any reason to be here and that moment when you can face the stars without a drink, when you can take in the beauty*

*and swallow the peace, then God takes you away and you become
dead because there is no longer a reason for you to be on the street
in the city* and O'Shay thinks to himself, *I gotta catch the magic,
my whole life has been trying to catch the magic.*

BOBBIE FINGERS THE TUMOR growing at the base of her neck.
She smiles. She has just bought a new puppy. Life goes on, her
smile says.

The smoke slowly curls out of her mouth. She says, "O'Shay
and I don't have lives. It's like chapters in a book. You finish one
chapter and you start another. Or you are sitting there with
nothing. Absolutely nothing."

She starts puppy-training classes next week.

The tumor grows, she smiles, the new chapter begins.

She remembers this lunch. Joey O'Shay has just taken down
a big heroin operation. He is so relaxed sitting there over his
plate of food.

And she is afraid to be at the same table with him because
she knows that the organization he just gored must have people
with guns looking for him.

The man who paced in that room for three months and
hardly said a word, the man she is certain will kill Joey, she
knows he is back in the picture. That no matter what Joey
O'Shay says about quitting, that man is back and waiting. And
he is the one.

Joey cannot read that man. He has told her so.

He is the one.

THE PHONE RINGS. Joey answers. The voice is that man who sat in
the room for three months and did not leave and spoke only to
Cosima. The man Bobbie says will be the one to kill Joey.

The man says, "You come to New York and I'll get you twenty-
five keys."

Joey refuses to come to New York.

The man says he will come to the city if necessary but then it will only be ten keys.

Joey says, It's up to you.

He hangs up.

He thinks if he comes here, if he brings that fine heroin, well, then the man has made his own decision.

But Joey thinks he will do nothing to bring about the deal. He will merely let it happen or not happen.

He thinks, I used to be able to do fifteen rounds. Now I can only do three.

He looks up, and on the office wall Angel Sonrisa smiles down from a photograph. Candles burn, the coffin waits, the altar is ready.

You can leave your life behind. If you have never had a life.

HE WILL PICK UP HIS BRUSHES. He can see the painting that says his life. A man is on the edge of a mountain. It is night and blackness coats the earth like tar. The man looks up. He sees the stars. And the man knows he will never understand what they mean. And he will never stop looking up at them in the black of the night, looking with wonder.

In deep night, there are no stars, just the lights and screams of the city.

But in the painting, they're out there every night.

He thinks, We are made out of stardust.

Stars.

I came to believe there was no escape
And there's gonna be pain, it's something you have to take
But I'm starting to find that there's cracks in these walls
And after the fall, there's love after all

—Ray Wylie Hubbard and Terry Ware, "After the Fall"

A Few Last Words

YEARS AGO, I WORKED on a newspaper in a hot desert city and the iron rule for photographers was never take a photograph of a cop with his hat off because the chief could not abide such slovenly ways. Naturally, given the heat, cops took their hats off and, naturally, photographers were careful to avoid capturing such moments lest the wrath and rules of a paramilitary culture fall on some patrolman's head. There is a touch of that decorum in this book and I'll spell it out.

Joey O'Shay cannot work alone, no one can—save a serial killer. He is embedded in a raft of technicians and other cops and agents and they help him in his work. They are largely cut out of this book so as not to disturb their careers. Just as some rough moments and talk about cops are included so as to be true to the frustration and stress that comes with the work. So you have a loner who is also part of a group and a man who in the end can never shed his identity as a cop. Whatever the friction of his life, the police world is the one to which he has given his years and his very being. In his closet, he keeps his unworn uniform clean and pressed, the blue blazing in the half-light of the room. The drawer opens in the bedside table by the four-poster and there in a plastic bag gleam all his medals.

The rest is as it came down. I was either there or talked to Joey O'Shay or got long e-mails in which he put down in words the heat of the moment. And the chill of the moment. I want to thank him for this privilege and thank a bunch of other people who trusted me. And who out of simple decency shall go name-less here.

I'd spent close to a decade in the drug world when I met Joey O'Shay. I wanted out. O'Shay had never talked to the press and pretty much held them and their magazines and newspapers in contempt. For reasons that don't matter here, we became friends. He asked one thing of me: that I read Viktor Frankl's *Man's Search for Meaning*. I expected some kind of crap about easing stress. I was wrong and I am grateful that he made this request of me. Joey O'Shay became my ticket out of this sewer I had been living in because his knowledge and his story and the les-sons he'd learn, well, they covered what I'd learned and had not said. And then some.

What I learned in the drug world is that eventually you vio-late yourself by selling out your own values. That you get mean, you get hard, you lie and there are those moments when the killing comes. Or your silence about the killings. You enter a world that others deny even exists and if you stay there, you either go down hard or you thrive. Some of us thrive, God help us.

But even at the very bottom of your personal arc, you re-member your former self and your better nature. For me this book is not about the drug world, it just takes place in that world. And it is about learning about one's own humanity as this very precious essence drains from one's veins. And this is not a book about cops, it just happens to take place inside the mind of a cop who called himself Joey O'Shay.

When the book was over, I went back to that unnamed city for one last long talk into the dark of night with Joey O'Shay.

I had no reason beyond my own need. He took me to a federal courtroom in a rural county where a twenty-nine year-old who had run an Ecstasy, coke, and methamphetamine ring went down hard for 361 months. O'Shay had been called in, though the case was not his, to elicit a confession. Which he did—with no clemency attached. His face remained blank as all but a life sentence came down. Then we had lunch with the prosecutor who had nailed the coffin shut on the defendant. She was a pleasant professional woman of thirty-five who regretted not slapping down more time on the culprit. I felt angry and dirty and yet at peace as I ate my hamburger and hand-cut French fries. The system eats people and the defendant was a lowlife killer who had met his doom in a rural, legal lynching party.

Afterward, we drove almost silently to Paradise Cove where O'Shay had fed his soul since childhood. We walked the shore as he explained that the defendant would almost certainly go to one of the federal super maxs, the new hells designed to lock a man up for twenty-three of the twenty-four hours. And that he would more than likely do the first ten years in isolation in order to break him to the wheel of justice. He stopped and picked up a hard, clay-colored rock and handed it to me. A white fossil, some shelled creature from the long ago, was embedded in the stone. He ran his future through his mind, *I'll float on the lake under the stars. I will beat this thing...it is my strategy...to win what I once was back. The Evil Creature that followed me will find and protect another...it is forever...A man as we are here...is not.*

We stayed there for some hours and then at dusk drove back to the city.

We sat in his backyard under the trees.

He drank beer.

I opened a bottle of red wine.

That is when he said it had taken him a thousand years,

a hundred lives to get to this moment, this place, this pain, this understanding.

It all made perfect sense to me.

That is what this book is about.

Getting to that place.

Last Call for Joey O'Shay

March 2006

THE PHONE CALL COMES AFTER the boy dies the third time. A hummingbird hovers a few feet away, near the orange fire of the honeysuckle. The light falls golden as October dismisses the furnace heat of summer. I sink into a chair on the patio and wait for the worst.

We haven't talked in a few weeks. I'd been meaning to call but somehow got tangled up with work and travel.

There'd been a hurricane down in Louisiana, and I'd gone down to write about it. I'd heard traces of Joey O'Shay in all the Cajuns explaining things to me, the roll of language ("Speak French? Been talking French for two hundred and fifty fucking years"). And I'd thought of him as the bayous flashed by in the night and I pushed the hot car up to around a hundred, since after the flood the cops were busy with other matters. Power was out everywhere and the stars reclaimed the sky.

I'd thought of him in London, too, where I went to write up a chef who served tripe. For days I'd dined on guts, wandered the lanes and enjoyed hating the British upper classes, the palaces, the litter of empty churches. I kept thinking my people had been sensible to get the hell out of this place. And that Tom Paine was right.

So I've gotten behind in my calling him. But I know a call at eleven on a Monday morning is not right. And from the first sound of his voice I know something has gone very bad. I've only heard this voice a few times before, and then it has been a fleeting thing—an adult man breaking down in the voice of a four-year-old boy as he remembers the death of his one-year-old son. I settle into the chair as he softly begins the inventory of moments. How on Thursday his twenty-four-year-old son, Skipper, felt under the weather, probably a touch of the flu. How on Friday Joey's mother had this feeling and called the older boy and told him to get over to Skipper's place pronto to check up on him, how he did just that and found his brother unconscious on the floor. How he watched Skipper die in his arms, but then, being a fireman and a paramedic, brought him back. And here that choking begins, that sobbing sound, soft but still a sobbing, and I mumble into the phone and say, Yes, yes, I understand, slow down, take it easy.

And then he recovers, the voice is under control again, and the inventory resumes. The older boy got the ambulance and called Joey at work—he was out on a homicide—and they met at the hospital, where the two of them watched Skipper die twice more on the table, and be brought back again each time. Now he's in a coma. Myocarditis, that's what did it. The heart is ravaged. The voice breaks again, there is a long pause. I make sounds of comfort, fragments of words that do not so much mean something as express something, and I say, Well, be in that room, you can't ever tell what a person can sense, I've spent long nights in such rooms. And he says yes, he knows, and we go on with soft words and silences and there is so much to say and we both know we cannot say it, not really, and that anyway there is little point in saying it, because we both know. And he wonders if this is payback for the way he has lived and I say, No, no, this is not payback, life is not a simple accounting system. This is not fair because life is not fair, but life is good and his son in a coma in a hospital, fighting for his life, surely knows that.

Oh, he says, that is so true. He tells me Skipper had met a girl, a nice girl, a nurse from New Orleans, and he'd brought her around and said, Dad, this is the one. He'd never seen his son this way about a girl, and everything had been so fine as summer ebbed and young love flooded the air. And maybe that was why he'd turned to his wife one night and told her that he was worried, that he'd always feared Skipper was too good and would be taken from him, and now he had that feeling again. Then the moment passed and it was once again summer and there was love floating around. Even last night, when they'd all come from the hospital, he'd turned to his older boy and said, Let's go out to eat, hell, call Skipper and see . . . and then he'd caught himself.

"I'm in denial, I guess," he tells me.

But he can't keep this thread going, his voice starts to break up once more. I feel helpless and start making those sounds again, the air rank with honeysuckle and the whir of hummingbirds around me feeding, and suddenly he says in a clear, firm voice, "Anyone asks you, you tell them that Kim Sanders is Joey O'Shay. You got that? Kim Sanders is Joey O'Shay."

And I know we've crossed over.

THERE ARE WAVES OF ENERGY, too big to control, too common to avoid, and when these waves come through, survival means slipping between them or riding them if you can, but it never means conquering them or besting them. The car is run off the road by armed men. The room is full of menace and guns and money and powder on the table. The street is dark and there is no one to call when trouble comes. You are asked to do something that is wrong and evil, but if you fail to do it you lose your job or your savings, and if you comply, you lose your soul. In these moments, you either survive or do not survive, but always, always,

you find out who you are and this discovery is not often pleasant. The words that are tossed around by others—heroism and
bravery and so forth—are lies, bald-faced lies to anyone who has
negotiated the waves and lived.

I thought Joey O'Shay had gotten home before dark. Two
months before this book was published, he'd left narcotics and
gone into homicide, believing that because murder was wrong
this work would not corrode his sense of self the way busting
drug dealers had.

There were good moments. Once, on a homicide, a young
cop sidled up to Kim and asked, "You're the guy, right? The guy
in the book?" And Kim said, "Yes." Then the young cop said,
"That Frankl guy, he knew, he really knew." And Kim felt very
good. A distant cousin in Arkansas stumbled onto the book and
called, and they spent a weekend catfishing and remembering
the scents of childhood.

Kim's sons were full-grown now, launched in their lives. Joey,
the older, had been a local news talent with an offer to move up in
his field when 9/11 happened. He quit, cashed out, and financed
himself through the training to be a fireman and paramedic,
work he enjoyed. Skipper was angling toward some kind of career in medicine. He'd been a football player in college, but despite his size he was a sweet kid, with an almost angelic glow about
him. I used to tell his father it was good the boy was huge and
strong because otherwise someone would catch that decency in
his eyes and mug him before he'd managed to walk a block. I remember sitting with the two of them at the dining room table and
the old man demanding that his son marry a big-boned girl because he wanted his grandchildren to be football players. Skipper
sat there under this barrage with an angelic grin, and they looked
more like brothers than father and son.

The book allowed the boys to fill in the pages of their
childhood, revealing to them the place where their father had

vanished for years. They caught the sense of their long-dead brother, and could finally speak to their father of this loss.

The only bad moments for them were reviews of the book that questioned its truthfulness, that treated it as some fiction. Names had been changed, of course, and places left unnamed in order not to jeopardize careers. Yet beyond these changes and silences everything was exact, thanks to surveillance records, e-mails, and the four years I'd spent hanging around the people and the deal. The boys knew as much, especially because they had lived through many of the things I wrote about. And because they knew their father.

Now Skipper lay in a coma with his ravaged heart, and Kim had crossed over. I understood that he no longer cared about hiding his identity or the city where he was based or mounting the slightest barrier of protection around himself. Things that had mattered a few days before meant nothing to him now.

DEATH HAD ALWAYS BEDEVILED KIM. He'd walk through the police academy where they'd hung the portraits of all the cops who had gone down in the line of duty, and he'd think, I know them, I've brushed against these people and now they're gone and where the hell are they. He kept the death of Angel Sonrisa in his mind by hanging a portrait so that whenever he looked up from work in his secret narc lair, the face of his old mentor and partner stared back at him, the sound of his laughter echoed, and his murder gave one more jolt.

But the stars kept coming, came back when he told me he was leaving drugs, that the deal in the book really was the last deal, because he no longer believed in what he'd been doing and that if he failed to stop doing it he was certain he would die. He drifted off into homicide, a new country where he had much to learn. And this new country felt clean, did not require setting up deals and betraying people. He'd even had a public radio crew out for two days, let them tape him talking as they'd taped one of his partners, took them around Dallas and introduced them to everyone, spent time at the lake and in the woods with them. And then it had been broadcast nationally, though without real names or an actual location given, and the roof on his life had not caved in.

He'd narrowed down his star quest to a county, a failed place where there were damn few people and not much going on, but the land was cheap and the sky at night was clotted with stars. Skipper had told him after he'd read this book that his star quest, his yearning for the feeling he'd had as a boy floating out at night in the jon boat, was "a longing and journey to return to innocence." He knew the kid was right.

But he could not really leave his life for the stars. Katrina ripped through Louisiana, and Kim had helped handle thirty thousand evacuees in Dallas. Then he'd gone home and watched as federal officials on television "congratulated and patted themselves on the back at a pulpit . . . before they even did shit. It reminded me of the lying-ass drug meetings and back-patting bureaucrats lying about what a great job they were doing on the drug war. Sorry fucking SOBs."

There were mopping-up details. The boy called Johnny Boy in this book came to trial and was sentenced to life in the Texas system, under the guidelines a minimum of forty years before he could even surface for a parole hearing. "He'll be sent to a supermax hellhole. One hour exercise in a cage. He had no reaction. He is dead, no one makes it to age seventy-four in a supermax hellhole." Johnny Boy's accomplice fell mute because he feared being murdered if he talked and so he went down for twenty. And then Kim mentioned another guy who had testified against Johnny Boy at the trial—the two of them had "glared at each other like poisoned pit bulls" and then the witness had "put it on him . . . He did it mainly because he trusted me, he told the DA."

I'd met the father of this witness, Kim told me, on under-cover patrol one afternoon, when Kim "slid him money by his daughter's apartment." And it all came flooding back to me, the late-day heat, the faces at the car door whispering tips, getting money, asking favors, the soft murmur of the drug world. This all happened back in the mists, I'm not even sure this book was in the works then, I was just hanging out in Dallas as Kim was showing me the world he knew and that he wanted me to know, a world that swallowed me whole for more than four years. I re-membered the young woman leaning in, anxious about her most recent urine test and her probation officer—"That bitch's gonna violate me," and Kim saying softly, "Give me her name, I'll make a call, it'll be okay," and I realized this was the lubricant that made the streets hum for both the cops and their prey. And I sat there wondering how I'd get this down, this tang of actual life, since I was dealing with a cop who did not want a story and with a world that demanded anonymity. And then I'd puzzled it out and written this book.

Now it was early September, the hurricane had taken an American city, Kim's boy, Skipper, was deep in love—he'd thought he was in love before, with that fine girl, but this was dif-ferent, his dad could tell—and the book was out and bubbling through Dallas. "People here know it is me," Kim continued, "and are buying it. Those who have read it so far have shyly told me it says things they felt . . . but could not mention or express. Some have said they did not know I'd lost a son . . . or painted . . . or even that Martha was my wife. I guess I was very private . . . for a very long time . . . and I'm glad. Some look like they want me to say something . . . but can't look me in the eye."

THE SECOND CALL COMES ON THURSDAY. He tells me it is hopeless, the doctors have made that clear. His boy is going to drown in his own toxicity. So they have suggested turning off the life-support machines. The voice begins to quaver, then there is a pause and the voice is back and matter-of-fact, and the voice says Skipper could linger a day, maybe two, after the machines are silenced.

Then he breaks.

He tells me that he went out on his front lawn, stood there with his wife and his mother before they went to the hospital and signed off on this final decision. The hedge had suddenly gone into bloom, odd for this time of year, but the weather had been strange lately, and there was a cloud of monarch butterflies, maybe twenty of them, swirling around the blooms. The three of them stood there in the sunlight and watched the display. Then Kim moved into the swirl and let the orange brilliance of life thrum around him. He put out a finger and a single monarch landed on it, then it fluttered up to his shoulder and stayed there, minute after minute. He froze, the two women stayed still, the rest of the butterflies swirled around him.

Then the one on his shoulder lifted off and did not rejoin the others but spiraled slowly up into the sky and finally disappeared into the blue.

"I know it was Skipper," he says, "I know he was saying goodbye. Martha and my mother both saw it. We knew."

I tell him he is right. That the butterfly is not a metaphor or symbol, that it is exactly as he saw and felt. I can hear him crying.

He hangs up.

I wonder why I told him that business about the butterflies. And I wonder why I have no doubts about what I told him.

THE MONARCHS LEAVE CANADA in August and head toward Mexico. Their two- to three-thousand-mile migration east of the Rocky Mountains is part of the parade of high summer in the United States. Eventually, the monarchs reach the mountains in Michoacan where they will winter over, millions clustering in blazing abundance. Come March they flutter north, lay eggs in the southern reaches of the United States, and then die. Their children and grandchildren and great-grandchildren continue to migrate north, and come August the wheel turns again. And as they flutter south to central Mexico, some people are delighted by the spectacle, some oblivious as they pursue their own journeys and rounds, and some, like Kim, transfixed in their front yards as they pause and stare before they head to the hospital and turn off the systems that support the life of a son. No one understands the monarchs' journey, no one knows how separate generations seem able to maintain it. For the moment, it simply is, something we see and yet cannot explain.

THE SLIPPAGE HAD BEGUN almost a year before. I would get calls and Kim would say, This is it, I'm going to pull my badge, I can't do this anymore. Then he would hesitate, and days and weeks would pass and then a similar call would come again. He went to Costa Rica with his wife and Skipper and Skipper's childhood friend Penny, and while he enjoyed the waters, he was haunted by the case and by the destruction of the people he had lured into the case, especially the woman called Gloria, the heroin broker who seemed to resist all the easy categories cops use to justify their work.

His long e-mails conveyed not the details of the case so much as his night thoughts about the meanings of the people involved and what the justice system had done to them, a kind of dark theology flying above and beyond the statements of the courts.

That was the time of the tortoise for me—a tortoise that lived in the yard, ravaged the flowerbeds, and seemed innocent of anything beyond sun for basking and plants for dinner. I left for Thanksgiving, and when I returned a strange dog had somehow gotten into the yard, and Daphne was a bloody blob of meat and chewed-up shell and visibly beating heart sitting on the path in the lower garden, seeking warmth from the early-morning sun. In a glance, I heard a *yes* forming in my mind, a tolling of some belief in life, and this led to a winter of antibiotics, weekends in intensive care, and finally reconstructive surgery. These matters were details compared to the hunger for life I felt in the small and violated body. Some part of me thought, If she refuses to give up, I cannot give up. There was a meaning to things beyond my own concerns and fears, and this creature knew the meaning without ever thinking of it for a moment, while I had to struggle even to notice that there was a meaning.

After my phone call with Kim, his voice breaking as he spoke of his dying son and the butterfly that had lingered on his body and then spiraled up and vanished into the blue sky, after he told me he knew what this meant and I agreed as to its meaning, these other thoughts clustered around, this sense of things, and so I stared out, godless and believing, stricken in my heart and yet sure of my faithless faith, and wound up telling Kim the butterfly was not metaphor but actual, and that I, too, felt his son in this brightly colored insect fluttering in his front yard. And this feeling did not leave but lingered and lingered.

ON FRIDAY HE CALLS AGAIN and this time the voice is different, no longer the fragile thing that broke down in the face of great pain and the inability to alter the reality of that pain. This time the voice is firm, the voice of Joey O'Shay in the room, doing a deal and in command. He tells me he went to the hospital and stood around his boy with a priest and family members, and suddenly Skipper opened his eyes, and his eyes were tracking. The boy was weak, so weak his hand could barely clasp another hand, and he could not speak, but for some seconds he looked and then he closed his eyes again and Kim stepped forward and peeled open one lid and the eye again tracked around the room. He was there, his boy was still there. And so Kim is again a creature of hope, and talks about what might be done when Skipper emerges from his coma and an inventory of his body can be taken, of heart transplants and other matters.

He says he's reached a decision. He is not going back to work. He has made a simple decision: "I'm going to live the rest of my life, each day, to honor my son." He is not sure exactly what this means or dictates, but he is certain how it will feel—that each of his days will add something good to life, will improve things. He is done with the way he has spent so much of his life.

He says that Skipper came to him after he had read this book and told him to put down this life, to leave this police thing, that he had been working since age fourteen and it was time for him to do something that made him feel good.

"Look," Kim continues, "he's not just my son. He's my best friend."

When he left narcotics, he took down the voodoo altar in his office and brought it home. Now he destroys it. And prays for forgiveness.

IN AN EARLY DRAFT OF THIS BOOK, there was a happy ending of sorts. Joey O'Shay and I had been talking for hours and then I'd turned off the tape machine and said, "That's it, we're done." And we each got a drink and went outside and left the book behind us for good.

O'Shay had drifted into a memory of a boy. He was always partial to tales of children, in books or movies or life—that innocence, and the pain of watching innocence end.

He said, "That tape's off, right?"

And I said, "Yes."

And then he began his story.

There is this boy and he is a natural athlete. His father is gone, his mother does this and that to get by. Joey O'Shay comes down to the ball field. One of his sons is playing. This is the Little League championship and there is a huge crowd.

In the last inning, his boy's team is down by three runs. They are facing a team from the monied part of town, a team that has fine uniforms and big players and has trounced every opponent. A storm is moving in, angry fists of clouds. O'Shay sits in the bleachers, watching. One kid comes to the plate and dribbles a single. Another gets a walk. His son hits a single.

The bases are loaded.

The winning run comes to the plate. There are two outs.

The boy at the plate is the kid with the father who's gone and the mother who scrapes by. He is a real shy kid. No one ever comes to watch him play.

The storm nears, O'Shay can feel its violence almost breathing down his neck. He looks at the boy, a boy he has tried to mentor. But it is hard to give hope to someone from such a home.

There is lightning, coming closer.

The kid stands at the plate and O'Shay can tell that he is terrified.

O'Shay feels himself praying, he's telling God, "You don't let this fucking game get called because of rain. You can take my ass to hell if you let that kid hit a fucking home run."

The sky is rumbling.

The kid takes two pitches, both strikes.

The other team is talking shit, they can feel victory.

The kid steps back from the plate and looks at O'Shay in the bleachers.

And O'Shay says, "You can do it."

And he knows the kid can understand him even with the thunder and the other team yelling shit and the crowd bellowing. He can feel his words reach the kid's face.

The kid knocks the ball out of the park.

They win by one run, and as the last runner crosses the plate, thunder, lightning, sheets of blinding rain. The kid looks over at O'Shay. They are alone in a screaming crowd as the storm roars in.

O'Shay sits and remembers that game, his face a mask, the eyes private as always, tears running down his face.

SUNDAY BEFORE NOON, the voice on the phone says, "Skipper's gone, he died an hour and fifteen minutes ago. His brother and I were there holding him," and then the voice breaks and sobs and repeats over and over, "Just tell them Kim Sanders is Joey O'Shay, just tell them."

A single monarch had lingered in the back of the house, right by the French doors where Joe the Crow came visiting, where the squirrels demand food. It'd been there for days, fluttering around whenever Kim looked out. And then that Sunday when he returned from his boy's deathbed, the monarch was gone.

AN HOUR AND A HALF AFTER THAT CALL, I begin the thousand-mile drive to Dallas. The Sonoran desert gives way to the Chihuahuan, the mountains frame the blue sky and the sun glows with the yellow of fall after the white light of summer. I do not play the radio or listen to music. For years I have driven this road, and I know the roll of the land. The mountains of west Texas sputter out to the south and then I swing north onto the Staked Plains, the homeland of the Comanches and after them the oil players and the roughnecks. I pass a prison at Big Spring where I sometimes visit a friend, slide down from the high plains into Sweetwater and then Abilene. I enter the tentacles of the sprawl that embeds Fort Worth and Dallas.

A single death seems small in the face of the land. A life is a brief thing, given the clatter of plants and animals and people along the road. I think that I lack the words for what I feel. I want to say that a life matters and I want to say that all lives matter and I want to comfort the afflicted. So I fall silent and seem more at home with birds in the sky, leaves fluttering the lazy breath of early fall, than I am with ideas or theologies or words of any kind. I open the *Dallas Morning News* and the obituary greets me.

A boy named Skipper played high school football and won the "Football Team's Knockout Award for Outstanding Offensive Play" at the final banquet. He joined the power-lifting team, made National Honor Society and the math club, also the fellowship of Christian athletes. Then came college, more football, a bachelor's degree in biology the past spring and notions of a life in medicine. In early October Skipper was felled by myocarditis, and on October 16 he "passed into heaven." He knew Spanish, loved diving in Costa Rica, and had a lot of friends. It rolls on this way, the stubborn effort of the living to make the dead into words, and then suddenly the tone changes. "He was always of a different place. He was of stardust and knew what most of us struggle to attain. He returned to his Father to be rocked gently. He dwells in a place of peace beyond human imagination."

After Skipper died, his grandmother answered her phone, and a voice that sounded very distant said, "Honey, I love you, I'm fine, I love you with all my heart." When the call ended abruptly, she thought the signal was simply breaking up. The old woman figured it was her surviving grandson, Joey, so she called him. But it was not. The boys had always sounded exactly the same. Kim tells me this and falls silent.

THE CHURCH FILLS AND STILL the line reaches out into the Dallas heat. A thousand people fill the sanctuary—family, friends, cops, firemen, lawyers. A mosaic frieze displays the stations of the cross, and up at the altar three huge photographs capture the boy, the football player, and the young man. I glance at the program and see that Kim is slated to speak, and I worry. Then I look over and see him organizing the procession of family members, his face blank, and I realize that Joey O'Shay is once again in charge. A bagpipe plays, a soprano sings "Ave Maria," the priest gives a prayer and a homily, three friends of the boy offer their thoughts. The father sits in some kind of trance. He calls on the spirit of his son to help him speak and speak honestly. Then he stands and begins: "My boy Skipper...he was different."

For fifteen minutes Kim speaks without faltering, the voice level and calm, the words floating over the congregation, audible and yet somehow hard to grasp.

"I look into this church," the voice says softly, "and I see so many loving friends and family. I look into this church and I see some hard men...I will tell you what Skipper wants you to know. There is great dignity in what you do though it can take you to the edge. You see terrible sadness and suffering. Use this hurt to be kind. Use this pain as a lesson and a learning process to a greater good."

Something about how he will never believe his son's life was meaningless, fragments about what Skipper was like and what he meant, and the words keep coming and coming—"He wants you to take the time to take your children away from the lights of the city. Let them have silence and gaze at the beauty of our heavens. Let them feel the magnitude of the universe and feel the warmth of eternity."

There is never the slightest hesitation or the smallest concession to grief.

"He knew," the voice finally says, "we come from stardust and long to return. It is the blue sky that makes your heart ache. It is the feel and smell of rain coming. It is the beautiful sound of waves on a beach in the moonlight. It is the smell of a child's hair.

"Father, rock my Skipper gently."

Communion is taken, the bagpipe plays again, the family files out.

Afterward men come up to Kim and say they could not imagine speaking so clearly at the funeral of a son.

And Kim is baffled.

He asks, "What man cannot speak for his child?"

YOU CAN NEVER BE WEAK. You learn this coming up and you keep relearning it on the streets. During that holiday in Costa Rica with his family, that strange and lonely time when Kim was supposed to be on vacation but kept thinking of all the people he had destroyed in his work, he felt very weak. He had sensed his life ebbing away for some time, believed he was sinking under a toxic wave of his own creation.

He took a day charter on a fifty-two foot sailboat with a captain and crew and they anchored in blue water off a perfect beach. So he plunged into the sea and decided to swim to shore.

With a few strokes, he knows he has made a mistake. His body cramps up, he begins to sink, and he thinks he will never see his sons again or his wife but will disappear into this deep empty blue.

He feels a huge hand grip him and pull him to the surface.

And Skipper starts swimming and towing his father to shore.

After a while, Kim worries that his son will tire and he will drown his boy.

He says, "You gotta let me go if you tire. Skipper, how long can you tow me?"

"Forever," he says. "Forever."

They make it to the beach and Kim is so weak he can barely walk. Kim is ashamed of his weakness. His son tells him they will keep what happened a secret.

The boy finds a sea cave and they sit down.

He pats his father on the back.

Kim says, "I love you as if I had known you forever."

Skipper says, "It is forever, Dad. The ocean. This cave. Us."

———

WE SIT ON THE PORCH. It is afternoon, the wake at the Irish pub is memory, as is the mass. The backyard is almost cleared of debris left by an informal wake of Skipper's friends that consumed several days.

Kim runs through all the manifestations of myocarditis, he's encyclopedic on the subject—eighty-four viruses, and only one need penetrate to wreak havoc. He's convinced that somewhere there is a door and some way he will reach that door and his son will be on the other side or some sense of his son. Out there, in the stardust. We talk for hours, all afternoon, slowly rocking in the white chairs on the porch as his son is cremated.

I tell him I don't know, I wish I knew, but I don't know.

But I know that life is not meaningless and do not doubt this fact.

He nods and sips iced tea.

Joey O'Shay is gone now, maybe for good.

He says, "When Skipper passed, electricity coursed through the room. And the nurse and heart doctor cried."

The nurse said later, "My God, you could feel it like a whirlwind."

He says, again, "I am going to live each day to honor my son."

He now thinks this: "I will be a better man . . . a kinder man. Even though I seem at times to feel as though he will walk through my door with those big arms and that big grin. I know I will someday walk through his. That is the place we all seek. It is where we came from . . . and what we need to go back to. I know that."

I leave at dusk.

In my rearview mirror, I see Kim out by the hedge, a few monarchs still swirling around the bloom. In the fading light, I see him lean forward, cup his hand, and watch a butterfly swirl down.

Kim Sanders, 1990

Acknowledgments

REBECCA SALETAN EDITED THIS BOOK from stem to stern, traveled a long way to meet Joey O'Shay, and shared in its odd evolution.

I want to thank Mary Martha Miles, who went over various drafts and helped clean up the mess I tend to create.

I wrote this book in Marfa, Texas, in a house loaned to me by a foundation as part of their writers' program. This is the second book I have written in that house thanks to the kindness of the outfit. So I would like to thank the folks at the Lannan Foundation. And issue an apology to the town of Marfa for once again being a hermit in their midst and remaining largely ignorant of their fine community. It seems when one is blessed with the loan of a Lannan house, one enters and hardly ever leaves the walls again. I did see things out my window that make me wish this were not my fate. The Davis Mountains and Big Bend region of Texas are good earth. Well, better than good.